Astrophysics: A Very Short Introduction

VERY SHORT INTRODUCTIONS are for anyone wanting a stimulating and accessible way into a new subject. They are written by experts, and have been translated into more than 45 different languages.

The series began in 1995, and now covers a wide variety of topics in every discipline. The VSI library now contains over 500 volumes—a Very Short Introduction to everything from Psychology and Philosophy of Science to American History and Relativity—and continues to grow in every subject area.

Titles in the series include the following:

ACCOUNTING Christopher Nobes
ADOLESCENCE Peter K. Smith
ADVERTISING Winston Fletcher
AFRICAN AMERICAN RELIGION
    Eddie S. Glaude Jr
AFRICAN HISTORY John Parker and
    Richard Rathbone
AFRICAN RELIGIONS
    Jacob K. Olupona
AGEING Nancy A. Pachana
AGNOSTICISM Robin Le Poidevin
AGRICULTURE Paul Brassley and
    Richard Soffe
ALEXANDER THE GREAT
    Hugh Bowden
ALGEBRA Peter M. Higgins
AMERICAN HISTORY Paul S. Boyer
AMERICAN IMMIGRATION
    David A. Gerber
AMERICAN LEGAL HISTORY
    G. Edward White
AMERICAN POLITICAL
    HISTORY Donald Critchlow
AMERICAN POLITICAL PARTIES
    AND ELECTIONS L. Sandy Maisel
AMERICAN POLITICS
    Richard M. Valelly
THE AMERICAN
    PRESIDENCY Charles O. Jones
THE AMERICAN REVOLUTION
    Robert J. Allison
AMERICAN SLAVERY
    Heather Andrea Williams
THE AMERICAN WEST Stephen Aron

AMERICAN WOMEN'S HISTORY
    Susan Ware
ANAESTHESIA Aidan O'Donnell
ANARCHISM Colin Ward
ANCIENT ASSYRIA Karen Radner
ANCIENT EGYPT Ian Shaw
ANCIENT EGYPTIAN ART AND
    ARCHITECTURE Christina Riggs
ANCIENT GREECE Paul Cartledge
THE ANCIENT NEAR EAST
    Amanda H. Podany
ANCIENT PHILOSOPHY Julia Annas
ANCIENT WARFARE Harry Sidebottom
ANGELS David Albert Jones
ANGLICANISM Mark Chapman
THE ANGLO-SAXON AGE
    John Blair
ANIMAL BEHAVIOUR
    Tristram D. Wyatt
THE ANIMAL KINGDOM
    Peter Holland
ANIMAL RIGHTS David DeGrazia
THE ANTARCTIC Klaus Dodds
ANTISEMITISM Steven Beller
ANXIETY Daniel Freeman and
    Jason Freeman
THE APOCRYPHAL GOSPELS
    Paul Foster
ARCHAEOLOGY Paul Bahn
ARCHITECTURE Andrew Ballantyne
ARISTOCRACY William Doyle
ARISTOTLE Jonathan Barnes
ART HISTORY Dana Arnold
ART THEORY Cynthia Freeland

James Binney

ASTROPHYSICS

A Very Short Introduction

OXFORD
UNIVERSITY PRESS

Great Clarendon Street, Oxford, OX2 6DP,
United Kingdom

Oxford University Press is a department of the University of Oxford.
It furthers the University's objective of excellence in research, scholarship,
and education by publishing worldwide. Oxford is a registered trade mark of
Oxford University Press in the UK and in certain other countries

© James Binney 2016

The moral rights of the author have been asserted

First edition published in 2016

Published in the United States of America by Oxford University Press
198 Madison Avenue, New York, NY 10016, United States of America

British Library Cataloguing in Publication Data
Data available

Library of Congress Control Number: 2015953131

ISBN 978-0-19-875285-1

Printed and bound by CPI Group (UK) Ltd, Croydon, CR0 4YY

Contents

List of illustrations

Astrophysics

List of illustrations

Chapter 1
Big ideas

In heaven as on Earth

Before Newton there was astronomy but not astrophysics.
If legend is to be believed, astrophysics was born when Newton
saw an apple drop in his Woolthorpe orchard and had the
electrifying insight that the moon falls just like that apple. That is,
a celestial body such as the moon does not glide on a divinely
prescribed path through the heavens as Newton's predecessors
supposed, but is subject to the same physical laws as the humble
apple, which tomorrow will be a half-eaten windfall not worth
picking up.

The power of this insight is that it allows us to apply physical laws
developed in our laboratories to understand objects that exist on
the far side of the universe. Thus Newton's insight enables us to
travel in the mind across the inconceivable vastness of the universe
to view a massive black hole at the centre of a distant galaxy from
which radio telescopes have received faint signals.

Newton laid the foundations of astrophysics in another key
regard: he showed that it is possible to obtain precise quantitative
predictions from appropriately defined physical laws. Thus he did
not only give a coherent physical explanation of observations that
had already been made, but he *predicted* the results of

observations that could be made in the future. To do this, he had to invent new mathematics—the infinitesimal calculus—and use its language to encapsulate physical laws. Since Newton's time most physical laws have taken the form of differential equations. A differential equation specifies a function by stating the rate at which it changes. The differential equation encapsulates what is universal about a given physical situation, and the initial conditions that are required to recover the function encapsulate what is particular to a specific event. For example the trajectory of a shell fired from a gun is the solution of Newton's equation $m \, d\mathbf{v}/dt = \mathbf{F}$, which is commonly abbreviated to $f = ma$ and relates the rate of change of velocity ($\mathbf{v}$) (the acceleration) to the force ($\mathbf{F}$) that is acting. Newton's equation applies to all shells and all falling apples, and to the moon. It is universal. The trajectories of the moon, the shell, and the apple differ by virtue of their initial conditions: the moon starts far from the Earth's centre and is moving exceedingly fast; the shell starts from the Earth's surface and is moving more slowly; and the apple also starts from near the Earth's surface but is initially stationary. These different initial conditions applied to one universal equation give rise to three completely different trajectories. In this way the mathematics that Newton invented became the means by which we identify what diverse events have in common and also what sets them apart.

It must hang together

James Clerk Maxwell, the only son of a prosperous Edinburgh attorney, early on displayed a great talent for mathematics and physics, and he made major contributions to the theory of gases and heat, and to the dynamics of Saturn's rings, but his greatest achievement was to extend the laws of electromagnetism by pure thought. He imagined a particular experimental set-up in which a alternating current flows in a circuit that includes a capacitor—a device consisting of two metal plates separated by a thin layer of insulator, which could in principle be a layer of vacuum. The current flows into one plate, charging it positively, and out of the other plate, charging it negatively. Maxwell applied to this circuit the rules for calculating the magnetic field generated by a circuit that had been developed by André-Marie Ampere. In 1865 he showed that these rules gave rise to completely different answers depending on how you applied them unless there was a current flowing between the plates of the capacitor, through the insulator. This result led Maxwell to hypothesise that a time-varying electric field generates a 'displacement current'. Mathematically, the hypothetical displacement current constituted an extra term in the differential equation that related a conventional current to the magnetic field that it generated.

The astonishing implication of the extra term in the equation was that it enabled the electric and magnetic fields to sustain each other without the participation of charges—until then an electric field was what surrounded a charged body and a magnetic field was what surrounded a current-carrying wire. But with the extra term a time-varying electric field generated a time-varying magnetic field, and Michael Faraday had already demonstrated that such a magnetic field generated a time-varying electric field. Thus the magnetic field regenerated the original electric field, without any charges being present! Could this amazing conclusion be correct, or was the extra term in the equation a foolish mistake?

Maxwell could calculate the speed at which the coupled oscillations of electric and magnetic fields would propagate through empty space, and that speed agreed to within the experimental errors with the measured speed of light. Maxwell concluded that his extra term was correct and that light *was* precisely mutually sustaining oscillations of the electric and magnetic fields. Because the wavelength of light was known to be extremely short (about 0.0005 mm) the frequency of the oscillations must be extremely high. Oscillations at lower frequencies would be associated with waves of longer wavelengths. In 1886 Heinrich Hertz generated and detected such 'radio' waves.

So Maxwell re-interpreted an old phenomenon, light, and predicted the existence of a completely new phenomenon by applying the conventional laws of physics to a thought experiment and arguing that the laws needed to be modified to ensure *consistency* of the theory. This was a ground-breaking step.

For ever and ever

We believe that the laws of physics have always been true: we have strong evidence that they were true a minute or so after the universe began 13.8 Gyr (gigayears) ago. They remained true as the universe evolved from exploding fireball through a cold, dark era to give birth to the first stars and galaxies, which are now being studied with huge telescopes. And they remain true to the present day.

Although the laws of physics have held steady over the last 13.8 Gyr, the universe has changed beyond recognition. Here again we have the Newtonian distinction between the laws of physics, embodied in differential equations, which are always and everywhere true, and the phenomena that they describe, which can change completely because the initial conditions for which we must solve the equations change radically.

Since the laws of physics are valid in every part of the universe, we can travel in our minds to distant galaxies. Because the laws of physics are valid at all times, we can travel in our minds back to the very beginning. The universal and eternal nature of the laws of physics enables us to become, in our imaginations, space-time travellers.

Astrophysics is the application of the laws of physics to everything that lies outside our planet. As such it is the child of other sciences but completely dwarfs them in its scope.

In the beginning was the Word

The universe is transitory, while the laws of physics are eternal. They were there before the universe started, and they structured the universe. The running of any particular experiment cannot be the same from day to day, because in the real world things change. Today it's colder than yesterday, and this fact *will* change how the experiment runs to some extent. The Earth's magnetic field is constantly changing direction, and this will affect the experiment to some extent. The Sun is growing older and increasing in luminosity, the moon is drifting way from planet Earth, and these facts will affect the experiment to some extent. In the real world nothing stays the same, but in the world of a physicist's mind there are laws that are eternally true, that never change. This fixedness isn't an accident and it isn't a mirage: it's an act of will. A physicist doesn't feel s/he understands a phenomenon properly until it has been traced back to a law that's eternally true.

If we pack all our equipment up and ship it to another country, to another latitude, the experiment will run differently, to some extent, because at the new location the Earth's magnetic field will be different, because the Earth's gravitational field will be different, because it will be hotter or colder, and the flux of cosmic rays through the laboratory will be different. But the laws of physics will be precisely the same. Again the sameness of the laws of physics here there and everywhere is an act of will: we will not rest until any difference between the way the experiment runs in the new location and in the old can be traced to some difference in the circumstances changing the solution we require to the immutable and universal laws of physics.

This insistence on explaining phenomena in terms of laws that are everywhere and always true doesn't only enable us to travel through space and time across the universe and back to the remotest times. It also equips us with three powerful weapons to take with us on our travels. These weapons are called energy, momentum, and angular momentum.

In 1915 Emmy Noether proved a crucial result. If the laws that govern a system's dynamics stay the same when the system is moved, or rotated, then as it moves or spins there is a quantity you can evaluate from its current position and velocity that will remain constant. We say the system has a 'conserved quantity'. The conserved quantity arising because the laws are the same everywhere is momentum, and the conserved quantity that arises because the system is indifferent to whether it is oriented east–west, north–south, or whatever other direction, is angular momentum. An extension of Noether's theorem is that if the dynamics is the same at all times then there is another conserved quantity, energy. Thus the universal and eternal nature of the laws of physics gives rise to three important conserved quantities, momentum, angular momentum, and energy. The constancy of these quantities is a big help when we are trying to understand a system that might be far away or back in the remote past.

In 1930 Wolfgang Pauli conjectured the existence of particles he called *neutrinos* that carried momentum and energy away during nuclear reactions. This conjecture was Pauli's reaction to experimental evidence that clearly showed non-conservation of energy and momentum. He conjectured the existence of unseen particles that ensure that energy and momentum *are* conserved. For a generation neutrinos were a pure speculation, but in 1956 they were finally detected. They are hard to detect because they have exceedingly small cross-sections $\sim 10^{-46}$ m^2 (metres squared), for interacting with anything. In the language of classical physics this means that a neutrino will collide with another particle only if it passes within $\sim \sqrt{10^{-46}}$ m $= 10^{-23}$ m of the centre of that particle, a distance that is 100 million times smaller than the size of a proton. Actually quantum mechanics makes it meaningless to localize particles so precisely, so the real implication of the very small cross sections of neutrinos is that they have a very small probability of interacting at all. Nonetheless, neutrinos play a significant role in structuring the Universe.

More happens in heaven than on Earth

Our story started with Newton bringing the moon down to Earth by subjecting it to the ordinary dynamical laws. In the 1930s the eccentric Swiss astronomer Fritz Zwicky restored the primacy of the heavens to some extent by asserting that 'if it can happen it will'. That is, *anything* that it permitted by the laws of physics will happen somewhere in the Universe, and with the right instruments and a bit of luck we can *see* it happening. Zwicky's Principle indicates that it is profitable to think hard about what weird objects and exotic events are in principle possible. If your knowledge of physics is good, you will be able to calculate what the observable manifestations of these objects or events would be, and perhaps even estimate how often they occur. Then you can encourage observers to look for these events.

The classic example of this process is the identification of white dwarf stars. In 1930 Subrahmanyan Chandrasekhar was taking the long voyage from Bombay to Southampton to work at Cambridge University. He wondered how the then new and controversial quantum mechanics might have implications for stars. He showed that when a star such as the Sun runs out of fuel, it will cool and shrink to a tiny volume—the Sun will in its time shrink to the size of the Earth—and the pressure that maintains this fantastically dense object from collapsing under the intense force of its gravity is a pure manifestation of quantum mechanics: even though the star is cool, its electrons will be whizzing about at near the speed of light because if they didn't move so fast, the least energetic of them would be violating Heisenberg's uncertainty principle, which requires an electron whose location is rather certain to have very uncertain momentum. Moreover, the Pauli exclusion principle forbids two electrons from occupying the same quantum state, so most electrons are obliged to occupy quite energetic states because the states that just avoid conflict with Heisenberg's principle are all occupied.

When Chandrasekhar reached Cambridge excited about his wonderful theory, he was devastated to have it dismissed as nonsense by the dominant figure of British astrophysics, Sir Arthur Eddington. Eddington didn't accept Zwicky's principle, and he didn't accept that quantum mechanics, a seriously flakey theory developed to explain (after a fashion) the behaviour of atoms, applied to whole stars. But Chandrasekhar was right, and there are quite close to the Sun large numbers of these cold, hyperdense stars, sustained by a pure quantum-mechanical effect.

In Chapters 3 and 8 we will encounter other examples of successful predictions of amazing things made with the creative use of physics. Zwicky's principle works because the universe is so huge and varied that nature has conducted within it a stupendous number of experiments. Our planet is a very interesting place, but a restricted one, and if you want to understand the material world, you have sometimes to look up and away from it.

A note on units

Standard scientific units, kilograms, metres, seconds, etc., are matched to everyday human experience and when used in astrophysics require some very large numbers. For us a more convenient unit of mass is the mass of the Sun, $M_\odot = 2.00 \times 10^{30}$ kg, where 10^{30} is shorthand for 1 followed by thirty zeros.

When considering planetary systems a convenient unit of length is the *astronomical unit* (AU), the mean distance of the Earth from the Sun: $1\,\text{AU} = 1.50 \times 10^{11}$ m, in other words, 150 million kilometres. On galactic or cosmological scales even an AU is too puny to be handy, and the unit of distance is the *parsec* (pc), which

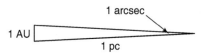

1. A parsec is the distance at which the Sun–Earth distance (1 AU) subtends an angle of 1 arcsec (1/3,600°).

is the distance at which a star that is stationary with respect to the Sun, when observed from the Earth, appears to move on the sky by one second of arc in a quarter year (Figure 1). From trigonometry $1\,\mathrm{pc} = 2.06 \times 10^5\,\mathrm{AU} = 3.09 \times 10^{16}\,\mathrm{m}$. The nearest stars lie about a parsec away, the centre of our Galaxy is $8.3 \times 10^3\,\mathrm{pc} = 8.3\,\mathrm{kpc}$ (kiloparsec) away and on average one galaxy as luminous as ours is contained in a volume of $\sim 10\,\mathrm{Mpc}^3$ (cubic megaparsec).

As our unit of time we usually take a year ($1\,\mathrm{yr} = 3.16 \times 10^7\,\mathrm{s}$) although we usually have to deal with longer timescales: stars evolve over millions or billions of years. Hence we often write Myr for megayears or Gyr for gigayears, where $1\,\mathrm{Gyr} = 1{,}000\,\mathrm{Myr} = 10^9\,\mathrm{yr}$.

A kilometre per second (km s) turns out to be a convenient unit of speed: the Earth orbits the Sun at $\sim 30\,\mathrm{km\,s^{-1}}$, and the Sun orbits the Galactic centre at $\sim 240\,\mathrm{km\,s^{-1}}$. Travelling at $1\,\mathrm{km\,s^{-1}}$ an object covers $\sim 1\,\mathrm{pc}$ in 1 Myr or 1 kpc in 1 Gyr. For example, in a gigayear the Sun covers $\sim 240\,\mathrm{kpc}$ while its path round the Galaxy has a length $2\pi \times 8.3\,\mathrm{kpc} = 52\,\mathrm{kpc}$, so it almost gets round five times in a gigayear.

The standard unit of power is the Watt (W) (roughly the rate of working when lifting a kilogram through 0.1 metres per second). A convenient astrophysical unit of power is the luminosity of the Sun $L_\odot = 3.85 \times 10^{26}\,\mathrm{W}$. Utilities generally charge for energy by the kilowatt hour, or $3.0 \times 10^{-28}\,L_\odot\,\mathrm{yr}$. A supernova explosion (Chapter 3, 'Exploding Stars') injects into the surrounding interstellar gas $\sim 8.2 \times 10^9\,L_\odot\,\mathrm{yr}$ of energy.

While $L_\odot$ yr is a convenient unit of energy for astronomical objects, it doesn't suit atoms at all. When discussing atoms and subatomic objects the convenient unit of energy is an electron volt—(eV). 1 eV is the energy required to move an electron through a potential difference of 1 volt and is $10^{-53} \sim L_\odot$ yr. The photons our eyes can detect each carry ~ 2 eV of energy, so in a year the Sun emits $\sim 10^{53}$ photons.

Chapter 2
Gas between the stars

The space between the stars is not completely empty although it is a vastly better vacuum than any that has been created on Earth: on average the gas near the Sun has 1 atom per cubic centimetre (cm^3) whereas air has $\sim 10^{19}$ atoms per cubic centimetre, so space near the Sun could be described as an ultra-high vacuum of 10^{-19} bar.

Interstellar absorption and reddening

This incredibly tenuous gas, mostly consisting of hydrogen and helium, manifests itself in many ways. One of the simplest and most important is through absorbing starlight. Actually, the starlight is not absorbed by the gas itself but by tiny particles of smoke that are embedded in the gas. Astronomers call these particles *dust grains*, but smoke is a much better name, for, as we shall see in Chapter 3, 'Life after the main sequence', they form in gases thrown off by certain stars precisely as soot forms in a burning candle or smoke forms in air drawn through a bonfire.

Naturally, the effectiveness with which dust absorbs starlight depends on the density of the dust, and therefore on the density of the gas within which it is embedded—it turns out that the mass of dust per unit mass of gas is roughly constant within our Galaxy. In a few directions the number of stars seen per unit area of the sky

2. **A dark globule.**

drops dramatically because in these directions there is a nearby dense cloud of gas, which obscures stars that lie behind it (Figure 2).

If you glimpse the Sun through the smoke of a bonfire, it appears redder than usual because blue light is more readily absorbed by small particles than red light. So red light from the Sun is more likely to pass through the smoke than the Sun's blue light. The physics of this selective absorption is that an antenna is an inefficient absorber of radiation with a wavelength much longer than itself: in the 1960s television aerials grew smaller when *ultra-high-frequency* broadcasting started (at $\sim 0.3\,\mathrm{GHz}$

12

(gigaHertz)), and mobile phones shrank and ceased to have visible antennae when it became cheap to make electronics that could process radiation with wavelengths $\sim$ 15 cm. It turns out that the vast majority of interstellar grains are smaller than one micron (10^{-3} mm), so waves with wavelengths longer than a few microns are not much absorbed by dust. In fact we can see right into dense interstellar clouds by observing at wavelengths of few microns, longer than the $\sim$ 0.5 micron wavelength of visible light by a factor of about 4.

Since dust grains are efficient absorbers of blue and ultraviolet light, stars seen through interstellar clouds look redder than similar stars that have little gas in front of them. By comparing the colours of such pairs of stars we can determine the *reddening* of the redder star and thus the amount of dust and therefore gas along our line of sight to the star. It was in this way that the existence of interstellar gas was first established.

Dust the regulator

Dust grains play a crucial role in regulating the temperature, density and chemical composition of the gas. Electrons and protons that are whizzing about in interstellar space sometimes bump into a dust grain. The force of the impact sets the dust grain oscillating, and these oscillations cause the grain to radiate electromagnetic waves. In this way some of the the kinetic energy of the electrons and protons is converted into electromagnetic waves, which, as we shall see, are likely to escape from even a dense gas cloud. It follows from this that dust grains are major coolants of interstellar gas.

We have seen that dust grains absorb plenty of starlight, especially blue and ultraviolet starlight. Naturally the grains are warmed by this absorption, just as a sunbather is. And because their masses are extremely small, the absorption of a single photon can raise a grain's temperature dramatically. That is, a single photon can set a grain quivering quite violently. Any electrons or protons that adhered to the grain after colliding with it before the photon was absorbed are then violently shaken off, somewhat as water is shaken off a dog who has just been swimming. If the electrons and protons that are shaken off the grain move away from it faster than they were moving when they banged into the grain, overall the grain will have heated the interstellar gas. Thus grains can cool or heat interstellar gas depending on the intensity of starlight in the gas.

If the starlight is feeble, several protons and electrons can accumulate on a single grain between absorptions of photons. Then the protons may come near enough to each other as they jiggle around on the grain's surface to form a molecule of molecular hydrogen (H_2). Energy is released during the formation of an H_2 molecule and this is donated to the grain. When a photon next warms the grain the H_2 molecule may float free. So dust provides the main mechanism by which atomic hydrogen becomes molecular hydrogen.

Grains broker many other marriages too. Interstellar gas contains carbon, nitrogen, oxygen, and sulphur atoms in much lower abundance than hydrogen or helium atoms, but in significant abundances nonetheless. If a grain has both carbon and oxygen atoms sticking to it, a molecule of carbon monoxide (CO) is liable to form. If a grain carries a carbon atom and a nitrogen atom, a molecule of an even more poisonous gas, hydrogen cyanide (HCN), is liable to form because there's usually a hydrogen atom around to make up the party. In this way dust grains control the chemical composition of interstellar gas.

Emission by gas

Much of what we know about interstellar gas has been gleaned by detecting radiation from interstellar atoms and molecules. Hydrogen atoms consist of an electron in orbit around a proton and emit detectable radiation in two very different ways. One mechanism involves the radiation flipping the spin of the atom's electron relative to the spin of its proton: the energy of the atom is slightly higher when the spins are anti-aligned than when they are aligned, so an atom with anti-aligned spins can emit a photon by flipping the electron's spin. The wavelength of this photon (21 cm) is very, very much longer than the size of the atom—so the atom is an incredibly inefficient radiator of this long-wavelength radiation. In fact, left alone, an atom with anti-aligned spins is likely to stay that way for over a hundred million years before collapsing into the lower energy state. Fortunately, it can be very peaceful in the backwaters of interstellar space, and there are phenomenal numbers of hydrogen atoms out there considering making the transition, so if you tune a radio antenna to the magic frequency there's a strong signal coming from flipping atoms. The existence of this signal was predicted by Frank van der Hulst while he was a student in Nazi-occupied Leiden in the Netherlands. Detecting this signal was made possible by war-time work on radar, and in 1951 its detection was announced simultaneously by groups in the Netherlands, Australia, and the USA.

The detection of 21 cm radiation from atomic hydrogen dramatically improved our understanding of our Galaxy because for the first time we could get a clear picture of the rotation of our Galaxy. The radiation betrays the state of motion of the emitting atoms because the frequency at which an atom radiates is very precisely specified for an observer at rest with respect to the atom. If the atom is moving with respect to the observer, the frequency measured is shifted, to higher frequencies if the atom is approaching the observer, or to lower frequencies if it is receding (the Doppler effect).

Hydrogen molecules don't interact with photons that have less energy than ultraviolet photons, and such photons are scarce. So H_2 is very nearly invisible. This is a major problem for astronomers because roughly half the interstellar gas in our

Galaxy comprises H_2, and in many ways it is the most important half because it is the cold, dense half that is liable to turn into stars and planets. Fortunately CO provides a good tracer of H_2. A CO molecule is an 'electric dipole' because the oxygen (O) atom grabs more than its fair share of the electrons, making the carbon (C) end of the molecule slightly positive and the O end negative. Unless the gas temperature is extremely low, the CO molecules spin as they move, and a spinning electric dipole emits electromagnetic waves. These waves are emitted at very precise frequencies because quantum mechanics restricts the possible spin rates to discrete values: the molecule can have no spin (quantum number $j = 0$) or one unit of spin ($j = 1$), or two units, and so on. Moreover, a molecule can change its spin state by only one unit at a time, and when it passes from spin state j to $j - 1$ it emits a photon that carries an amount of energy that is proportional to j. So all the frequencies of the photons emitted are multiples of the fundamental frequency associated with the transition $j = 1 \rightarrow j = 0$. These fundamental photons have wavelength 2.3 mm, and since wavelength is inversely proportional to frequency, the wavelength of the transition $j \rightarrow j - 1$ is $2.3/j$ mm.

The probability that a given molecule is spinning at the rate j depends on the temperature of the gas—if the temperature is low, there is not much energy around and few molecules are either spinning fast or moving fast, whereas at high temperatures the molecules tend to both spin and move fast. Consequently the ratio of the number of molecules in the state $j = 4$, say, to those in the state $j = 1$ increases with rising temperature. It follows that as the temperature rises, so does the intensity of the spectral line with wavelength $2.3/4$ mm relative to that with wavelength 2.3 mm. Hence measurements of several of these spectral lines enable us to determine the temperature of the gas.

In the 1970s it became possible to survey our Galaxy in the first few of these lines and thus to map the denser, colder part of the interstellar medium.

Nearby galaxies have long been mapped in both the 21 cm and 2.3 mm spectral lines. It is now possible to detect CO in very remote galaxies, and a global effort is underway to build a giant radio telescope, the Square Kilometre Array, which from 2020 will enable us to map the distribution of 21 cm emitting gas before the first stars and galaxies formed.

In its ground state atomic hydrogen does not absorb visible photons, but it does absorb energetic ultraviolet photons—those that carry more than 10.2 electron volts. Photons that carry 10.2 eV are called Lyman α photons and they play an important role in astronomy because they are very readily scattered by hydrogen atoms—the atom absorbs a photon and subsequently re-emits it in a different direction.

Photons that carry more than 13.6 eV of energy can strip the electron right out of a hydrogen atom. That is, they can ionize a hydrogen atom—convert it into a free electron and a proton. Subsequently, the proton is likely to capture a passing free electron, emitting a photon as it does so. The first photon emitted may carry only a small amount of energy because at first the electron may be only marginally bound. But once the electron has become trapped it is very likely to fall, like a drunkard who loses his/her balance on a staircase, deeper and deeper into the proton's electric field, emitting another photon as it slips down each step. Photons emitted as the electron falls to the next to lowest step on the staircase are known as *Balmer photons*. The least energetic of Balmer photons, Hα photons, have a beautiful pink colour and show up nicely in photographs of places where stars are forming because the hot stars in these regions are ionizing the gas around them, and thus preparing the ground for proton–electron recombinations.

Ultraviolet photons have a big impact on molecules as well as atoms because they can break molecules into their constituent atoms—*dissociate* molecules. In fact this is the main way in which molecules are destroyed—the molecules are formed on dust grains and destroyed by ultraviolet photons. Hence the chemical composition of interstellar gas hinges on the balance between the destructive power of ultraviolet photons and the catalytic action of dust grains. The higher the density of the gas, the more frequently atoms collide with dust grains and the larger is the proportion of atoms that are tied up in molecules. Moreover if the density of dust grains is high, the dust grains will absorb a significant fraction of the ultraviolet photons from hot stars before they can dissociate a molecule. So the fraction of the gas that is in molecular form increases rapidly with gas density.

If the density of gas in some region becomes high, a runaway situation can arise in which the density rises and rises almost without limit. This runaway occurs because as the density of the gas increases, fewer of the ultraviolet photons emitted by nearby stars penetrate deep into the cloud before being absorbed. But we have seen that grains are the major heat source of interstellar gas, and that molecules such as CO radiate energy. Hence a falling density of ultraviolet photons and a rising molecular fraction causes the gas to cool. Cooler gas exerts less pressure at the same density, so as it cools, the cloud contracts under the inward pull of its gravity. As the density rises, ultraviolet photons become still scarcer, the gas cools and contracts further. This runaway density leads to the formation of first the dark globules seen in Figure 2, and then stars.

The gas disc

Observations of the 21 cm line of atomic hydrogen and the 2.3 mm lines of CO reveal a thin layer of gas around the Galaxy's midplane. Beyond about 4 kpc (kiloparsec) from the centre, the gas is moving at close to the velocity of a circular orbit. The CO is more centrally concentrated than the atomic hydrogen, and more lumpy. It is also more concentrated towards the midplane, mostly being within

$\sim 40\,\mathrm{pc}$ (parsec) of the midplane rather than $\sim 100\,\mathrm{pc}$ for the atomic hydrogen.

Every so often interstellar matter is blasted by a huge explosion. We will describe the objects (supernovae) that blow up in Chapter 3, 'Exploding Stars'. Here we consider the impact these explosions have on interstellar gas.

A supernova ejects between one and several solar masses at speeds of several thousand kilometres a second. The kinetic energy of the ejected gas is $\sim 10^{44}\,\mathrm{J}$ (Joules). For comparison, over its 4.6 Gyr life the Sun has radiated less than $6 \times 10^{43}\,\mathrm{J}$, so we are talking about a serious quantity of energy.

The first gas to be ejected from the supernova slams into the essentially stationary ambient gas, compressing it, heating it, and jolting it into motion. Naturally, all this effort slows the ejected gas, so it is then hit from behind by gas that was ejected from the supernova a little later. This gas now slows, compresses, and heats. In this way a thick expanding shell of hot, compressed gas forms around the supernova. On both its inner and outer edges the shell is bordered by discontinuities in the gas velocity: on the outside it is slamming into stationary gas, and on the inside it is slowing gas streaming out of the supernova.

As the shell sweeps up more and more interstellar gas, it is cooled by expansion. If it expands undisturbed for long enough, its temperature will eventually drop to the point at which it is cooled by radiation faster than the two shocks are heating it. The denser the ambient gas is, the sooner this condition is reached because the luminosity of the shell is proportional to the product of its density and mass.

In Chapter 3, 'Star formation' we shall see that stars form in clusters, so the supernovae that mark the ends of the lives of massive stars cluster too. Hence a second supernova often goes off in the low-density region inside the expanding shell of an earlier supernova. Then the shell around the second supernova expands quickly through the low-density region and merges with the first supernova's expanding shell. Now we have a *supernova bubble*, which may well recruit further driving supernovae as it expands. In Orion there is a region of very active star formation with regular supernova explosions that is driving a wall of atomic hydrogen towards us at over $100 \, \mathrm{km \, s^{-1}}$—this wall is called *Orion's cloak*.

Some supernova-driven shells of fast-moving atomic hydrogen move away from the Galaxy's midplane and launch the gas onto an orbit in the Galaxy's gravitational field that carries the gas far from the plane. In fact roughly 10 per cent of the Galaxy's stock of atomic hydrogen is more than a kiloparsec away from the Galactic plane. Eventually this gas falls back to the plane, so one says that the supernovae are driving a *galactic fountain*.

When such a sheet of cool gas is shot into orbit, the way is clear for gas ejected by a supernova to flow clean out of the Galaxy. The gas is so hot that its electrons are rarely bound to ions, so it comprises free charged particles. In this condition we say a gas is a *plasma*. It is likely that this has been an important process over the life of the Galaxy and intergalactic space is rich in supernova ejecta.

If star-forming events are sufficiently common, adjacent bubbles will overlap. Even though supernova explosions are thought to occur in our Galaxy once in $\sim 50 \, \mathrm{yr}$ (years) on average, supernova bubbles have overlapped to the extent that most of interstellar space is filled by them, with the denser interstellar gas squeezed into the narrow spaces between bubbles. The pressure (P), density (n), and temperature (T) of an ideal gas are connected by Boyle's law, $P = \mathrm{constant} \times nT$, and the pressure exerted by the hot gas is roughly the same as that exerted by the cold gas, so the T and n of interstellar gas have an approximately constant product nT even though T varies from 20 Kelvin (K) to $2 \times 10^6 \mathrm{K}$.

Supernovae accelerate electrons and ions to relativistic energies (Chapter 6, 'Shocks and particle acceleration'). These particles stream along the magnetic field lines that lace all interstellar space. From time to time a relativistic ion bumps into a nucleus of the interstellar gas, creating a gamma ray. The rate at which this happens in any volume is roughly proportional to the density n of the gas, so the intensity of gamma-ray emission is a valuable way to probe the density of interstellar gas.

We have seen that stars have a big impact on interstellar gas. Now let's take a look at how stars work.

Chapter 3
Stars

To this day our most important probes of the universe are telescopes that gather either visible photons or photons with only slightly longer wavelengths (infrared photons). At these wavelengths the night sky is entirely dominated by stars. We detect about a billion stars individually, as tiny unresolved points of light, and a billion billion more as contributors to the light coming from galaxies so distant that we cannot distinguish individual stars in the great agglomerations of stars that galaxies are.

So most of what we know about the Universe has been gleaned from a study of stars, and one of the major achievements of 20th-century science was to understand how stars work, and to understand their life-cycles from birth to death.

Star formation

Stars form when a cloud of interstellar gas suffers a runaway of its central density as discussed at the end of Chapter 2. After the density has increased by an enormous factor, of an order of a million million (10^{12}), photons emitted by atoms and molecules start to have trouble escaping from the cloud because they are scattered by molecules and dust grains after going only a small distance through the dense mass of gas and dust. When you pump up your bicycle tyres, the pump becomes warm from the work you do compressing air inside it. Similarly, as gravity compresses the gas of a collapsing cloud, work is done on the gas and the gas will warm if it cannot radiate the newly acquired energy. Once photons find it difficult to escape from the cloud, the work done by compression cannot be radiated in a timely manner and the temperature begins to rise. However, even as the temperature and pressure rise at the centre of the cloud, the crushing force of gravity increases too as more and more gas falls onto the core of the cloud. The consequence is a prolonged period of rising central temperature and density. If a cloud is sufficiently massive, the temperature and density are eventually sufficient to ignite *nuclear burning*, energy release by transmuting hydrogen to helium, and then helium nuclei into heavier nuclei such as carbon, silicon and iron. We discuss nuclear burning below.

When an interstellar cloud suffers a runaway increase in its density, it does not form one star but a whole group of stars. We do not understand completely this process of fragmentation, but it is an important empirical fact. Within the deforming interstellar cloud several regions of runaway density arise, each capable of seeding a star. The rates at which these seeds accumulate mass varies greatly, with the result that a few give rise to massive stars and many give rise to low mass stars. The most massive stars have masses $\sim 80\,M_\odot$, and the masses of stars extend down to below the mass $\sim 0.01\,M_\odot$ at which a star is too faint to detect at any point in its life.

Since the original cloud was a heaving, swirling mass of gas, the seeds move with respect to one another. One aspect of this motion is that one seed will get in the way of gas falling onto another seed, so augmenting its own growth and suppressing that of its neighbour. Another aspect of the relative motion is that seeds often go into orbit one around another to form a binary star.

Elsewhere whole groups of seeds go into orbit around each other to form a gravitationally bound cluster of stars. In Chapter 7, 'Slow drift', we shall see, however, that small star clusters are not stable and tend to evolve into a binary and a series of single stars.

As the seeds accumulate mass and begin to resemble stars, their nuclear energy output becomes more and more significant for gas in the lower density parts of the original cloud. The more massive stars start to radiate ultraviolet photons, which heat low-density gas as we saw on page 13. In Chapter 4 we shall see that young stars are surrounded by bodies of orbiting gas called accretion discs, and that these discs eject jets of gas along their spin axes. These jets slam into and heat diffuse gas in the neighbourhood. The upshot of all this activity by young stars is that quite soon after the density in a cloud runs away at specific locations, most of the cloud's gas is heated up and driven away. Consequently, from a cloud containing, say, $10^4 \, M_\odot$ of gas, only $\sim 100 \, M_\odot$ of stars will form. This low *efficiency of star formation* enables galaxies like our own to go on forming stars at a fairly steady rate throughout the age of the Universe because it implies a low rate of conversion of interstellar gas into stars.

Nuclear fusion

Atoms consist of a tiny positively charged nucleus surrounded by one or more electrons that move on orbits that take them $\sim 10^{-10}$ m from the nucleus. Nearly all the atom's mass is contained in the nucleus, which is only $\sim 10^{-15}$ m across. When two atoms collide, the orbits of the electrons deform so the distribution of negative charge surrounding each nucleus changes, and the nuclei experience electrostatic forces that deflect them from their original straight-line trajectories. The upshot is that even a head-on collision of two atoms is unlikely to lead to a collision of the nuclei themselves because their velocities will be reversed by electrostatic forces before the nuclei have a chance to come into contact. In a sense an atom's electrons provide a sophisticated anti-shock packaging for the nucleus that carefully protects the nucleus in all but the most extreme collisions.

As the temperature rises at the centre of a forming star, the atoms whizz about faster and faster and the violence of their collisions steadily increases. Electrons are knocked clean out of atoms so more and more nuclei become bare. Still colliding nuclei are unlikely to come into physical contact because, as positively charged bodies, they repel one another electrostatically. But eventually the collisions are so violent that some colliding nuclei actually touch. At this point nuclear reactions start to take place.

The energy scale of nuclear reactions is a million times larger than that of the chemical reactions that power our bodies and our cars. So the release of energy by nuclear reactions in the core of a cloud is a game changer. The density is soon stabilized at the value at which nuclear reactions release energy at just the rate at which heat diffuses outwards through the now massive overlying envelope of gas; if the rate of energy release is slightly lower than the rate of outward leakage, the central pressure falls, the core collapses, the temperature and density rise, and so does the rate of nuclear reactions. Conversely, if nuclear reactions are releasing energy faster than it can diffuse outwards, the central pressure rises, the core expands, the temperature falls and so does the nuclear reaction rate. Thus nuclear energy release makes a star an inherently stable mechanism.

Key stellar masses

A star more massive than $0.08\,M_\odot$ now settles to the business of nuclear burning. Stars more massive than $0.08\,M_\odot$ but less massive than $\sim 0.5\,M_\odot$ burn hydrogen to helium but cannot ignite helium. Stars with initial masses in the range $0.5 - 8\,M_\odot$ burn hydrogen and then helium, but cannot ignite carbon. Stars initially more massive than $8\,M_\odot$ but less massive than $\sim 50\,M_\odot$ burn carbon to silicon and then silicon to iron. Iron nuclei are the most tightly bound so no energy can be obtained by transmuting iron into any other element—iron nuclei constitute nuclear ash.

Stars more massive than $50\,M_\odot$ become unstable and explode before they have reached the stage of silicon burning. We know they become unstable, but are not certain what the final outcomes of these instabilities are. We think most of the star's mass is ejected into interstellar space leaving only a black hole as marker of the star's existence.

Stars with initial mass smaller than $0.08\,M_\odot$ only get hot enough to burn deuterium to helium. Deuterium is an isotope of hydrogen in which the nucleus consists of a proton bound to a neutron rather than a lone proton. Deuterium like hydrogen was created in the Big Bang and is destroyed in stars. It is $\sim 10^{-5}$ times less abundant than ordinary hydrogen, so it does not take long for a star to exhaust this fuel. An object that is only burning deuterium is called a brown dwarf. When the deuterium is consumed the object will cool to become an almost undetectable black dwarf.

The main phase in the life of a star more massive than $0.08\,M_\odot$ is the burning of hydrogen in its core. Since three quarters of the original interstellar cloud comprised hydrogen, in this stage there is lots of fuel to burn, and, as a bonus, more energy per nucleon (a neutron or proton) is released when hydrogen is burnt than when any other nuclear fuel is burnt. The Sun has been burning hydrogen in its core for 4.6 Gyr and it is only half way through the process. For reasons that will become apparent when star clusters are discussed in Chapter 3, 'Testing the theory', we call a star that's burning the hydrogen in its core a *main-sequence* star (Figure 3).

The more massive a star is, the more quickly it depletes its stock of core hydrogen and the shorter its main-sequence lifetime. Massive stars are spendthrifts: the bigger the inheritance of fuel they have at birth, the sooner they are bankrupt by virtue of having consumed that fuel. Figure 3 quantifies this fact by showing the

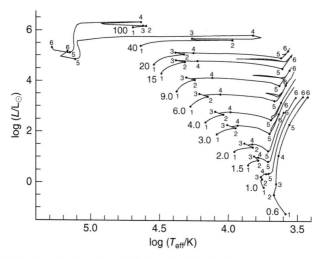

3. Luminosity plotted vertically in units of the luminosity of the sun as a function of surface temperature in degrees K for stars of various initial masses. The large number at the left end of each curves give the mass of a star in units of the solar mass $M_\odot$. While on the main sequence a star sits near the dot at the left end of its curve. It moves away from this dot when it has converted most of its core hydrogen to helium.

luminosities of stars of different masses as functions of surface temperature. During its main-sequence phase a star moves between the dot at the left end of its curve in this diagram and the point on the curve marked '2'. The big numeral gives the star's mass in solar masses. The vertical scale is logarithmic so there is a factor of a million in luminosity between the main-sequence points of stars with masses $0.6\,M_\odot$ and $20\,M_\odot$. Consequently, while a star of mass $0.6\,M_\odot$ will remain on the main sequence for 78 Gyr, nearly six times the age of the universe, a $20\,M_\odot$ star will be on the main sequence for just 8.5 Myr.

Figure 3 quantifies another key fact: the surface temperature of a main-sequence star increases with its mass. So massive stars are luminous, hot, and have short lives, while low-mass stars are faint, cool, and have long lives. When you heat a piece of metal strongly, it first glows dull red, then becomes yellow and white, and if you could raise its temperature even further it would glow blue. Hence hot stars are blue while cool stars are red. Figure 3 shows that all blue stars are massive, and we have seen that massive stars have short lives. So blue stars are always young.

The link between the colours and temperatures of stars reflects an important piece of physics. A *black body* absorbs any photon that hits it and emits a characteristic spectrum of radiation that depends only on the body's temperature and not on what the body is made of. Radiation with this spectrum is called *black-body radiation*. To a first approximation, a star is a black body and emits black-body radiation at the temperature of its photosphere.

Life after the main sequence

Once the hydrogen in a star's core has been consumed, hydrogen burning occurs in a spherical shell around the helium core and the core grows more massive, contracts and becomes hotter. In Figure 3 the star moves rather rapidly from point 2 to point 6 on its track, and we see that during this manoeuvre the luminosity increases, while the surface temperature falls. The rise in luminosity is most dramatic for low-mass stars, which have low main-sequence luminosities, and the decline in surface temperature is most pronounced for massive stars, which have high main-sequence temperatures. In fact, stars that have reached point 6 on their tracks all have rather similar temperatures, 2,000 K.

The reason the surface temperature drops as the luminosity rises, is that the increased flux of nuclear energy flowing from the hydrogen-burning shell puffs the star's enveloping gas into a bloated heaving body of gas in which energy is less transported by outward diffusion of photons that by convection. Convection is the process by which radiators heat rooms: hot air that has been in contact with the radiator rises, and is replaced by cooler air that falls down surfaces such as window glass and external walls that are unusually cool. The restructuring of the envelope into a bloated convective mass enlarges the star's photosphere, the sphere that emits most of the star's light. The swollen photosphere can radiate even the increased luminosity at a lower temperature than before. Since main-sequence stars are smaller than they will become once they have depleted their core hydrogen, they are called *dwarf stars*, and they will evolve into *red giant* stars.

At the point 6 in Figure 3 the helium in the star's core ignites. In stars more massive that $\sim 2\,M_\odot$ the ignition is quiescent, but in less massive stars the helium ignites explosively and we speak of the *helium flash*. The energy released in the flash causes the star's envelope to pulse violently and a significant portion of the envelope is ejected back to interstellar space. In stars more massive than $2\,M_\odot$ the regulatory mechanism described on page 25 functions properly and helium burning starts quietly. The onset of core helium burning causes the luminosity to drop slightly and the star to become slightly bluer.

The period of a star's life during which it quietly burns its core helium is the second most extensive period after that of the main sequence. Stars with initial masses up to $2.5\,M_\odot$ all spend ~ 130 Myr in this phase. At higher masses the duration of this period decreases rapidly with mass, and for a $20\,M_\odot$ star it is a mere 0.6 Myr.

Once the core helium has been burnt, helium burning shifts to a shell around the carbon core, beyond which hydrogen burning continues is an outer shell, and the star's luminosity rises rapidly. The star's envelope swells and instabilities frequently cause significant parts of it to be ejected into interstellar space. During this period these stars blow away most of their original mass in an increasingly powerful wind. As the gas flows outwards it cools and elements that form solids with high melting points condense into dust grains, so these stars have much in common with a Victorian factory chimney.

The rate of blow-off becomes a runaway process because as the star grows less massive, the power of gravity to hold gas in against radiation pressure in the envelope diminishes. Moreover, the luminosity of the star, and therefore radiation pressure grows as the quantity of gas that is blanketing the intensely hot core diminishes—the star's loft insulation is disappearing. Eventually the bottom of the envelope, where helium burning was taking place, lifts right off and the envelope becomes an expanding shell of gas around the star that is powerfully illuminated by the now naked core. The shell is ionized by photons from the core and glows brightly—the object is now a *planetary nebula* (Figure 4).

In the core nuclear reactions have ceased, so it gradually cools. It has become one of the white dwarf stars whose physics Chandrasekhar correctly outlined on the voyage from Bombay (page 7).

Stars initially more massive than 8 $M_\odot$ ignite carbon, burn it to silicon and then ignite that and burn it to iron. Since iron cannot be burnt, they are now obliged to replace heat that leaks out of their cores by contracting and thus releasing gravitational energy. Unfortunately, when a self-gravitating body contracts, its central temperature rises, and this rise in temperature soon proves fatal for the star—it suddenly blazes up into fireball and becomes a supernova.

The surfaces of stars

The density of gas in a star decreases continuously from the centre outwards, at first gradually but with increasing speed, although the density never falls precisely to zero. As the density falls, the distance a typical photon can travel before it's scattered or

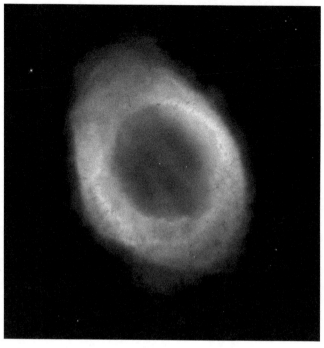

4. The planetary nebula Messier 57.

absorbed by an atom increases. At a certain radius this distance quite suddenly becomes comparable to the distance over which the density halves, and many photons can escape from that radius to infinity. The observed properties of the star are largely determined by the physical conditions in the spherical shell of this radius, the *photosphere* (Figure 5).

Photons of different frequencies escape from the star from radii that increase with the photon's propensity to be scattered by free electrons. Some photons have an unusually high propensity to scatter because they resonate with an oscillation of a common atom or molecule, and these remain trapped to the largest radii.

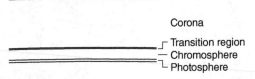

5. The outer layers of the Sun. Sunlight comes from the photosphere. The temperature of the solar plasma reaches a minimum in the lower part of the chromosphere. In the transition region the temperature leaps from ~ 10,000 K to over a million degrees. The blisteringly hot corona extends very far out and is readily observed during a total solar eclipse.

Hence the brightness of the star varies with wavelength and the star's spectrum contains spectral lines. The shape of these lines conveys information about radial gradients in density and temperature around the photosphere. Consequently, astronomers make huge efforts to obtain high-quality spectra for large numbers of stars. The precision with which the mass, radius, temperature and chemical composition of a star can be inferred from its spectrum is often limited by our ability to compute to the necessary accuracy the spectrum of light emitted by a star of particular mass, radius, etc.

Stellar coronae

The temperature of material in the Sun falls steadily all the way from the centre to the top of the photosphere—the visible surface—where it is about 4,500 K. Then, astonishingly, it starts to rise, at first slowly and then extremely rapidly—in the *transition region*, which is only 100 km thick, the temperature surges from ~ 10,000 K to over 1,000,000 K (Figure 5). Since heat always flows from hotter to cooler material, the corona must be heating the Sun, not the other way round. So what heats the corona? Outer space?

Convection carries much of the heat generated in the Sun's core on the last stage of its journey to the surface. Blobs of hot gas rise from 210,000 km below the photosphere, come to rest in the photosphere and there cool by radiating into space. Finally, they fall back to be reheated below the surface. Although convection is mainly an up-and-down process, in the photosphere gas does move horizontally before falling. So convection drives an unsteady circulation of gas.

The highly ionized gas in the Sun is a near perfect conductor of electricity because the many free electrons move easily in response to the tiniest electric field. Magnetic field lines freeze into and are swept along by such a conducting fluid, and the Sun's gas is magnetized. So the chaotic stirring of the Sun's surface by convection is constantly stretching and tangling the field lines that are embedded in the gas.

Magnetic field lines are analogous to elastic bands: there is a tension along the field line, and if a field line is stretched by the flow, the field grows stronger and its tension increases. In this case the fluid does work on the field; conversely, if the field line contracts, the field works on the fluid.

Adjacent field lines that are running in the same direction repel each other (Figure 6). If the field happens to be running parallel to the surface, this pressure pushes the field lines that are nearer the surface up and away from the field lines that run deeper down.

Gas cannot move across field lines, but it can flow down them, and once a field line has started to bow upwards, gas runs down the field line away from the crest of the bow. This flow diminishes the

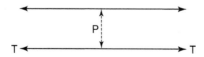

6. Each magnetic field line is under tension and repels similarly directed field lines.

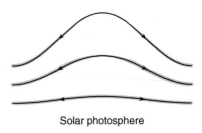

Solar photosphere

7. Plasma draining away from the crests of three upward bowing field lines.

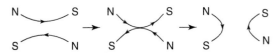

8. When magnetic field lines that are moving in opposite directions are brought together, their oppositely directed sections cancel out, releasing magnetic energy, and the field settles to a different, 'reconnected' pattern.

weight that is bearing on bowed field lines, so they rise up some more, encouraging more gas to drain away from the peak, and soon a big loop of magnetic field is sticking out of the Sun's surface (Figure 7). Meanwhile, the dense gas in which the two ends of the loop is embedded flows over the Sun's surface in response to both convection and the systematic rotation of the Sun, and the field lines that make up the loop often become tangled in the sense that field lines that are moving in quite different directions are dragged close to one another. At this point the field *reconnects* somewhere in the corona as sketched in Figure 8.

When field lines reconnect, energy stored in the magnetic field is used to accelerate particles. Most of these energetic particles collide with nearby electrons and ions and lose their extra energy by heating nearby gas. So in the vicinity of a reconnection event the gas becomes extremely hot. Thus the searing heat of the corona is maintained by a constant flow of magnetic energy from the turbulent layer that is bounded by the photosphere through the 2,000 km thick chromosphere, a region of low-density gas that envelops the photosphere (Figure 5).

Much of the gas in the corona is too hot to be confined by the Sun's gravitational field, so it flows away from the Sun as the *solar wind*. About 60,000 km from the Earth the wind is deflect around us by the Earth's magnetic field. Electrons that have been accelerated to extreme energies in reconnection events above the photosphere, escape into the wind without losing much of their energy, and some of these particles become trapped in the Earth's magnetic field. These particles make up the *van Allen* radiation belts. They race at close to the speed of light from one magnetic pole to the other, exciting air molecules to glow as they approach the surface of the Earth near the north pole—this is the origin of the northern lights.

Exploding stars

Very occasionally a single star will become for a week or two as luminous as a whole galaxy of 100 billion stars. Such an event is called a *supernova*. In a galaxy like ours we expect a supernova to occur roughly every fifty years, although the last Galactic supernova to be observed was that found by Johannes Kepler in 1604—remnants have been found of supernovae that exploded in about 1680 and 1868, but the events themselves passed unnoticed. In February 1987 a supernova, *SN1987a*, was observed in the Large Magellanic Cloud, a small galaxy that is currently passing very close to our Galaxy and will eventually be eaten by it. This event provided by far the best opportunity to observe a supernova that mankind has so far enjoyed.

Supernovae are key cosmological tools because they can be observed out to vast distances. Consequently, major observational resources have in recent years been devoted to detecting and measuring large numbers of supernovae.

It turns out that two completely different mechanisms can generate a supernova.

Core-collapse supernovae

In the incredibly dense cores of stars that have burnt carbon to silicon, much of the pressure that resists gravity is provided by the electrons, which are obliged by the Heisenberg and Pauli principles (page 7) to whizz about much faster than they would do at the same temperature but a lower density. As a consequence they have so much kinetic energy that it can be energetically advantageous for them to become trapped inside a nucleus, lowering its charge and thus transforming it into the nucleus of the element before it in the periodic table. Each such capture reduces the number of electrons that contribute to the pressure opposing gravity.

As the core contracts, its temperature rises and the average energy of the photons in the ambient black-body radiation (page 28) rises. Eventually, this radiation contains a significant number of photons that are energetic enough to blast an atomic nucleus to pieces (*photo-dissociate* it). The gas of photons in the core makes a significant contribution to the pressure that resists gravity and each photo-dissociation reduces the pressure by withdrawing energy from the photon gas.

Hence the star is on a slippery slope: contraction drives up the temperature, which leads to more electrons being captured and more photo-dissociation of nuclei, which inevitably lead to further contraction. Within a few milliseconds the core is in free fall and cataclysm is inevitable.

As the central density rises, the atomic nuclei, so patiently assembled through the life of the star, are blasted apart. Most of the fragments end up as neutrons. Now the neutrons start to play the role that was earlier played by the electrons: they make a major contribution to the pressure by moving much faster than they would at the same temperature and a lower density because they have to conform to the Heisenberg and Pauli principles. Consequently, at some point the pressure within the core rises steeply with density and the core abruptly stops contracting, or 'bounces'.

Most of the star's mass lies outside this pressure-supported core and is falling inwards very fast. The inevitable result is a shock (Chapter 6, 'Shocks and particle acceleration') in which the inward falling material is brought to rest and violently heated.

The temperature and density are now so high that collisions within the plasma of electrons, neutrons and protons generate neutrinos in abundance. Because neutrinos have incredibly small cross sections for colliding with anything (page 6), they move significant distances between collisions even in the stupendously dense centre of the star. As a consequence the core become enormously luminous by radiating neutrinos rather than photons—it radiates photons too, but they are slow to diffuse outwards, so the neutrinos carry off energy much faster. So at this stage an enormous flux of neutrinos is flowing out through the envelope of the star, most of which is still falling onto the almost point-like core. A small fraction of the neutrinos collide with infalling nuclei, transferring energy and momentum to them. These transfers can be sufficient to reverse the inward motion of much of the envelope and blast it outwards in a great ball of fire.

Before the material of the envelope disperses into interstellar space, it is exposed to an intense flux of neutrons boiled off the neutron-rich core. Nearly all the neutrons are absorbed by atomic nuclei in the envelope, converting them into heavier nuclei. Usually the nucleus formed by the absorption of a neutron is highly radioactive and quickly decays to another nucleus, often by the emission of an electron and always with the emission of a photon. Hence radioactive decay becomes a significant source of heat within the dispersing envelope. Some nuclei absorb several neutrons one after the other, and undergo several radioactive decays. All elements that lie beyond iron in the periodic table were formed in this way—thus the nuclei of bromine, silver, gold, iodine, lead, and uranium were all created in supernova explosions.

As the fireball expands, its photosphere swells, so its optical luminosity rises. In the case of SN1987a the optical luminosity peaked three months after the core imploded—we know when the latter happened because the blast of neutrinos as the core bounced was detected. (To date SN1987a is the only supernova from which we have detected neutrinos.) Spectra taken at this stage show the material of the envelope to be fleeing the core at a few thousand kilometres per second. At this speed each solar mass of ejected material has 2×10^{43} J of kinetic energy, so the $\sim 5\,M_\odot$ of ejected material contains $\sim 10^{44}$ J of energy. This is just ~ 1 per cent of the gravitational energy released by the collapse of the core. This is a characteristic feature of supernovae: the spectacular explosion we detect and the profound impact that the event has on the interstellar medium are both powered by a tiny fraction of the energy that is actually released: 99 per cent of the energy is carried off by neutrinos that will never interact with anything, ever.

Eventually the expanding envelope becomes so diffuse that optical photons cannot be trapped for long within most of it. Hence its optical luminosity fades over a few weeks. As it expands and becomes more diffuse, dynamical interaction with the gas that was in the volume around the star becomes more important. In fact the gas density in this region is likely to be anomalously high, because before the star imploded as a supernova it was blowing off mass quite fast as a wind. The blast wave from the exploding star ploughs into the wind and shock heats it. Significant emission at radio wavelengths can arise at this stage.

Meanwhile, back in the core, much has been happening. We have seen that the collapsing core became very neutron-rich, and pressure generated by the neutrons caused the core to bounce and generate a burst of neutrinos that ejected much of the envelope. After the bounce, material continues to fall onto the stabilized core and a neutron star takes shape: this is a stupendously big atomic nucleus that's seriously neutron-rich. Within it neutrons dash around at mildly relativistic speeds, creating the pressure that resists gravity in just the same way that electrons resist gravity inside a white dwarf. The mass of this neutron star grows as material continues to fall onto it. As its mass grows, its radius shrinks (white dwarfs behave the same way), its gravitational field grows yet stronger, and the neutrons move ever faster to resist gravity. Remarkably, general relativity predicts that pressure is itself a source of gravity, so the harder the pressure resists gravity, the stronger gravity becomes. If the mass of the neutron star increases beyond a critical value, M_{crit}, gravity overwhelms the pressure generated by the neutrons and the object collapses into a black hole.

The precise value of M_{crit} is controversial because it depends on how matter behaves at nuclear densities. Although we believe we know what equations we need to solve, those of *Quantum Chromodynamics* (*QCD*), it is extremely hard to determine from first principles the required relationship between density and pressure. Moreover, we do not have good access to the relevant regime experimentally—the most massive nuclei on Earth contain only ~ 240 protons and neutrons and generate negligible gravitational fields. The experts are confident, however, that M_{crit} lies between $1.4\,M_\odot$ and $3\,M_\odot$, so some core-collapse supernovae leave neutron stars, while others produce black holes. The supernova recorded by the Chinese in 1054 left a neutron star (the *Crab pulsar*) that has been extensively studied. Our understanding of SN1987a implies that it involved the formation of a neutron star, but meticulous searches have failed to reveal a relic star of any kind.

Deflagration supernovae

We saw above that stars with initial masses smaller than $8\,M_\odot$ fail to ignite carbon and lose their envelopes, leaving a core of carbon and oxygen that gradually cools as a white dwarf star. If the star has no companion, that is the end of its story. The majority of stars do have a companion, however, and then the future can be much more exciting. If the companion was initially the less massive star, its evolution will occur on a longer timescale. Consequently, the companion will swell up and start blowing off mass at a significant rate after its companion has become a white dwarf. If the distance between the two stars is not too great, a significant part of the mass blown off by the companion will be captured by the white dwarf's gravitational field and form an accretion disc. Accretion discs of this type are studied by their X-ray emission (Chapter 4, 'Time-domain astronomy'). At the inner edge of the disc, gas transfers from the accretion disc to the white dwarf star, so the latter's mass increases.

As the mass increases, the star's radius decreases and its gravitational field becomes more intense. The Heisenberg and Pauli principles then decree that the most energetic particles speed up. Eventually some nuclei are moving fast enough to trigger the conversion of carbon into silicon. This conversion releases energy, which heats the star, so more nuclear reactions take place.

If the thermal motion of the nuclei were making a significant contribution to the pressure within the star, the star would respond to the heat input from nuclear reactions by expanding and cooling, both of which changes would slow the rate of nuclear reactions, and the system would be stable. But in a white dwarf the nuclei make an insignificant contribution to the pressure, which is dominated by the electrons. Consequently the density does not decrease as the nuclei are heated, and the rate of nuclear reactions spirals out of control. The technical term for the way the rate of nuclear reactions runs away is *deflagration*, a kind of slow explosion in which a front of enhanced temperature and reaction rate moves through the medium at a speed that is slower than the speed of sound.

Within a fraction of a second about a solar mass of carbon and oxygen has been burnt all the way to iron and nickel. The sudden release of energy by all this burning *does* produce enough pressure to overwhelm gravity and the star flies apart, leaving no gravitationally bound object where it was. A significant part of the dispersed material of the star consists of a highly radioactive isotope of nickel (^{56}Ni) which has a half life of 6.1 days. Gamma rays produced by the decay of ^{56}Ni to iron heat the dispersing material and cause it to glow brightly. It is by this glow that we detect deflagration supernovae. They are generally called *type Ia SNe* from the empirical classification of their optical spectra.

Type Ia SNe are important for astronomy in two different ways. First it proves possible to estimate the luminosity of a type Ia SN from the rate at which its brightness declines. Consequently, these objects can be used as *standard candles*: objects of known luminosity whose distances can be determined from their apparent brightnesses. Second type Ia SNe are major producers of iron since nearly all the original white dwarf is eventually converted to iron. Core-collapse supernovae, by contrast, produce a cocktail of heavy elements that is much richer in the *alpha elements*, which include carbon, silicon, magnesium, and calcium. Consequently, by measuring the abundance in a star of iron relative to the alpha elements, one can determine the relative importance of type Ia and core-collapse supernovae to the enrichment of the star's material. In Chapter 7, 'Chemical evolution', we shall see the value of this determination.

Testing the theory

The theory of stellar evolution requires as inputs a great deal of atomic and nuclear physics and involves both extensive numerical calculations and some assumptions about how turbulent fluids mix. Can we be sure that it is correct? We think it is fundamentally sound because it has now been possible to compare several aspects of what it predicts to the actual outcomes of observations.

Globular star clusters

Globular star clusters provided the classic tests of the theory in its formative years. Our Galaxy has about 150 star clusters that are very nearly spherical and quite compact—in a cluster such as NGC 7006 (Figure 9) several tens of thousand stars lie within ~ 10 pc of the cluster centre—for comparison, within 10 pc of the Sun there are fewer than a hundred such stars. The feature of a globular

9. The globular cluster NGC 7006.

cluster that makes it an ideal test of the theory of stellar evolution is that, to an excellent approximation, its stars differ only in their masses: they have the same distance, age, and chemical composition. Moreover, the chemical composition of the stars can be estimated from the stars' spectra. For any conjectured age of the cluster and distance to it, the theory then predicts that the stars will lie on a curve called an *isochrone* in a *colour-magnitude* diagram, a plot such as Figure 10 in which the brightnesses of stars are plotted against their colour. For historical reasons, blue colours (implying hot surface temperatures) are plotted on the left of the diagram. The vertical brightness scale is always logarithmic, so changing the assumed distance to the cluster shifts the isochrone along which stars are predicted to lie up (for reduced distance) or down without distorting the isochrone's shape in any way. Changing the assumed age changes the isochrone's shape in computable ways. The age and distance are determined by finding the isochrone with the shape that best matches the observed distribution of stars and then finding the vertical position that produces the best match.

As Figure 10 illustrates, excellent matches between theory and observation can be produced in this way. Nevertheless, slight discrepancies do arise, and astronomers continue to refine the data and assumptions that go into stellar modelling to diminish these discrepancies. The ability of the theory to match data for many clusters leaves little doubt of its fundamental soundness, however.

A fascinating aspect of these fits is that our Galaxy's globular clusters prove to be extremely old—at one time the ages being derived were inconsistent with the age of the Universe. Since then refinements in our understanding of both the cosmic expansion and stellar evolution has yielded consistent ages. Clusters that have the lowest abundances of 'metals' (elements later in the periodic table than helium) tend to be the oldest, and even these have ages $\sim$ 12 Gyr that are not larger than the age on the Universe, 13.8 Gyr.

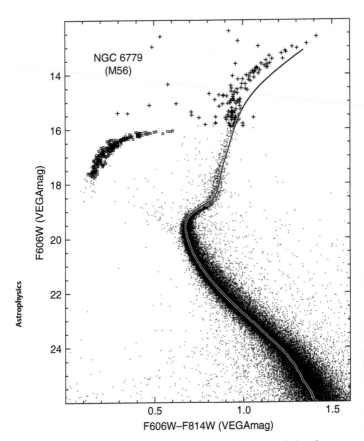

10. Brightness plotted against colour for stars in the globular cluster NGC 6779. Blue stars are on the left and bright stars at the top. The curves are theoretical isochrones (see text). The points from bottom right up to the sharp bend comprise the main sequence. Stars here are burning hydrogen in their cores.

Solar neutrinos

Every time two protons fuse in the Sun to make deuterium that shortly afterwards produces helium, a neutrino is produced that escapes from the Sun. Consequently, the Sun radiates neutrinos in addition to photons. In the 1960s Ray Davis set out to detect neutrinos from the Sun, arguing that their detection would be an important test of the theory of stellar evolution since they come to us direct from the Sun's energy-generating core and thus probe an entirely different region from the photosphere, from which we receive photons.

Davis's experiment involved processing tons of dry-cleaning fluid (tetrachloromethane CCl_4) down a mine—only by working in a mine could he exclude cosmic rays, which would generate a tiresome background signal. The CCl_4 was a convenient store of ^{37}Cl nuclei, which could become argon (Ar) after being hit by a neutrino. The argon had to be extracted and its quantity measured. After years of hard work, Davis measured only a third of the expected flux of neutrinos from the Sun. The scientific community was not very excited by Davis's failure: some wondered whether the experiment could detect Ar as efficiently as was claimed, others doubted the accuracy of the predicted neutrino flux.

These doubts were troubling because the experiment was sensitive to only a small minority of the neutrinos produced by the Sun: neutrinos produced by the fusion of protons don't have enough energy to transmute Cl into Ar; the more energetic neutrinos Davis hoped to detect came from other reactions whose rates are very temperature sensitive, and don't produce much of the Sun's energy. A small change in the rate at which heat is carried from the Sun's core could drastically lower the flux of these energetic neutrinos. So the experts re-examined their models of the Sun, but they were unable to reduce the flux of the higher energy neutrinos enough to be consistent with Davis's experiment.

Another issue was that Davis's experiment was only sensitive to one type of neutrino: there are three kinds of neutrinos, associated with electrons, muons, and tau particles. Davies's experiment could detect only electron neutrinos. This should not have been a problem as the nuclear reactions were expect to produce electron neutrinos. But could somehow two thirds of the emitted neutrinos pass through Davis's lab as undetectable muon or tau neutrinos?

From the mid-1980s two huge neutrino detectors were built that use either water (H_2O) or heavy water (D_2O) as the detector. A major advantage of these detectors is sensitivity to all three types of neutrino. One of these detectors, the *Kamiokande II detector* in Japan, observed the burst of neutrinos from SN1987a, and the other, the *SNO detector* in Canada, showed that when all three neutrino types are counted, the flux of neutrinos from the Sun is consistent with the original model predictions. This result from the SNO detector confirmed the idea that as a neutrino makes its way out of the Sun, it morphs from an electron neutrino into the other kinds of neutrino, with the result that roughly equal numbers of the neutrinos of each kind reach the Earth. Thus the astrophysics of the Sun had been correct from the outset, and the problem with Davis's experiment lay with particle physics.

The idea that neutrinos oscillate between the different types was first invoked to explain the outcome of solar-neutrino experiments, but later the process was studied in some detail using beams of neutrinos from nuclear reactors. It is a particularly important phenomenon because it implies that neutrinos have non-zero rest masses. Various experiments constrain the rest mass of the electron neutrino to a small value < 2 eV but neutrino oscillations establish that neutrinos have non-zero rest masses.

Stellar seismology

Stars like bells have natural frequencies at which they can oscillate. Each oscillation frequency is associated with a particular *mode* or type of oscillation. Bubbling associated with convection excites a star's modes of oscillation, and a great deal about the structure of a star can be learnt from measuring the spectrum of frequencies at which a star oscillates. Consequently, since about 1985 major observing programmes have monitored first the Sun and later nearby, relatively bright, stars to determine their oscillation spectra.

A star's modes are of two basic types. The easiest to appreciate are the *p-modes*. These are analogous to the modes of an organ pipe: a standing sound wave is set up that involves alternating compression and rarefaction of the air in the organ pipe or gas in the star. P-modes are of less interest astrophysically than the other type of mode, *g-modes*. These are a little like ocean waves: when a less dense fluid (air) sits on top of a denser fluid (water), waves in which the surface of the denser fluid oscillates above and below its equilibrium level can propagate over the interface between the two fluids. Waves on the surface of the ocean are associated with a discontinuity in fluid density, but within the body of the ocean there is often a continuous gradient in density associated with salinity: salty water is denser than fresh water so in equilibrium more saline water underlies less saline water, and waves that distort the contours of equal salinity move across the ocean.

In a star, gas that has lower *entropy* underlies gas of higher entropy. Entropy is a measure of the thermal disorder in a fluid; it is increased by conducting heat into the fluid and decreased by extracting heat from it. It is distinct from temperature in that when air is compressed in a bicycle pump or the cylinder of a diesel engine, the air's temperature rises, but its entropy stays the same. Waves in which the surfaces of constant entropy oscillate up and down propagate around the star in the same way that waves in the surfaces of constant salinity propagate through the ocean. G-modes are standing waves of this type.

Oil companies prospect for oil by launching seismic waves with explosions and detecting the waves at remote sensors. A computer subsequently deduces the density and elastic properties of the rocks in the region from how the waves have travelled through the Earth from the source to the detectors. The pattern of frequencies of a star's oscillation modes is similarly sensitive to the density and rotation velocity of gas at different levels within the star, so with appropriate software one can constrain the density and rotation velocity within the star. These values can be compared with the predictions of theoretical models. The major uncertainty in the models is the star's age, which must in practice be estimated by matching models to the observational data, and the star's spectrum of normal modes most strongly constrains the age because as a star ages, its central concentration increases: the core contracts, growing hotter and denser, while the envelope expands. This evolution changes the pattern of oscillation frequencies.

The seismology of the Sun has demonstrated that models of the Sun that are founded on a wide range of data from nuclear and atomic physics work very well but not perfectly. Small discrepancies between the predictions of the models and the seismographic findings probably arise from limitations of the atomic data, or the stellar models. But it's just possible that they point to completely new physics being involved in the transport of energy out of stars: 'dark-matter' particles (Chapter 7) could become trapped inside stars and on account of their low propensity to scatter off other particles contribute disproportionately to the outward transport of heat.

Binary stars

At least half of all stars are members of a binary system, and the existence of binary stars is a major issue for the theory of stellar evolution because when the more massive, faster evolving, star swells up as it becomes a red giant star, its companion is liable to grab gas from the swelling envelope. This theft causes both stars to deviate from the evolutionary path we have described for single stars because mass is a key determinant of stellar evolution, and now each star's mass has become a function of time.

As material falls onto the more compact, less massive star, energy is radiated (Chapter 4). Some of this radiation heats the outer layers of the more massive star, and they may become hot enough to escape from the binary altogether as a wind.

Both the transfer of mass from one star to the other and loss of gas in a wind will change the binary's orbit and can draw the two stars closer together. If the stars do move closer, the rate of transfer or ejection of matter will accelerate, so these processes can run away and lead to the two stars merging. In fact the more massive star may envelop the less massive star in its swelling envelope even if the binary's orbit does not evolve up to that point.

Once the less massive star is inside the more massive star, its orbital motion will be opposed by friction. The envelope will be heated and the less massive star will spiral inwards. After a time the system will have become a single star, but neither the star's core nor its envelope could have been produced by the evolution of a single star.

In short, close binary stars constitute a Pandora's box of complexities, and attempts to understand the contents of this box are an active area of research.

Chapter 4
Accretion

When you empty the kitchen sink after washing up, the water usually swirls around the plug hole and leaves a column of air in the centre of the waste pipe as it runs away. This swirling action arises because the water has a tendency to conserve its angular momentum as it flows towards the waste pipe. Angular momentum per unit mass is given by the formula

$$L = rv_t,$$

where r is the distance from the point around which the fluid is flowing (the centre of the waste pipe) and v_t is the speed of the fluid's motion perpendicular to the direction to the centre. As the fluid flows inwards, r decreases, and v_t has to increase to keep L constant. Tornados (twisters) provide a more dramatic example of the same physics in action: as air is drawn towards the tornado's centre to replace warm, damp air that has floated up to high altitude, the air spins around the twister's centre faster and faster until it is moving so fast it can rip roofs off buildings, pick up cars, and generally wreak havoc.

The principle of angular momentum conservation was crucial for the creation of the disc of the Milky Way, within which the Sun is located, and in the formation of the solar system itself. In this chapter we will see that it plays a large role in many of the most exotic and luminous objects in the Universe.

Accretion discs

In all these systems gravity is sucking gas in towards some centre of attraction, which may be the centre of a galaxy, a star, or a black hole. That is, these objects are *accreting* gas, and conservation of angular momentum causes the accreting gas to spin around the accreting body as it moves in. If the gas is cold in the sense that its pressure is insufficient to provide effective resistance to the inward pull of gravity, the spinning gas forms an *accretion disc* in which the gas at each radius is effectively on a circular orbit around the centre of attraction.

Basic disc dynamics

In thinking about the structure of an accretion disc it is helpful to imagine that it comprises a large number of solid rings, each of which spins as if each of its particles were in orbit around the central mass (Figure 11). The speed of a circular orbit of radius r around a compact mass such as the Sun or a black hole is proportional to $1/\sqrt{r}$, so the speed increases inwards. It follows that there is *shear* within an accretion disc: each rotating ring slides past the ring just outside it, and, in the presence of any friction or *viscosity* within the fluid, each ring twists or *torques* the ring just outside it in the direction of rotation, trying to get it to rotate faster.

Torque is to angular momentum what force is to linear momentum: the quantity that sets its rate of change. Just as Newton's laws yield that force is equal to rate of change of momentum, the rate of change of a body's angular momentum is equal to the torque on the body. Hence the existence of the torque from smaller rings to bigger rings implies an outward transport of

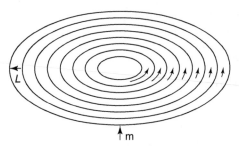

11. An accretion disc is imagined to comprise many solid annuli, all spinning at different rates about their common axis—the length of the curved arrows arrows is proportional to the local speed. The time it takes an annulus to complete a rotation increases outwards. Mass flows inwards and angular momentum *L* flows outwards through the disc.

angular momentum through the accretion disc. When the disc is in a steady state this outward transport of angular momentum by viscosity is balanced by an inward transport of angular momentum by gas as it spirals inwards through the disc, carrying its angular momentum with it.

As gas enters the outer radius of a ring, it has more energy than it has when it leaves the inner radius of the ring because its gravitational potential energy has decreased by twice the amount that its kinetic energy of rotation has increased. Hence each quantity of gas that passes through a ring deposits a quantity of energy in the ring. Moreover, the next ring in is doing work on our ring by torquing it in the direction of rotation at a rate that exceeds the rate at which our ring is working on the next ring out. Hence our ring also gains energy from the viscously driven flow of angular momentum through the disc. The energy gains from both the inward flow of matter and the outward flow of angular momentum heat the ring, causing its material to glow. This is why systems with accretion discs can be luminous astronomical objects.

To a good approximation each ring radiates as if it were a black body (page 28), so the spectrum of radiation from the accretion disc can be computed from the temperature $T(r)$ as a function of radius. The temperature of the ring of mean radius r is computed by balancing the rate at which energy is deposited against the rate at which radiation carries away energy. If the accreting body is gaining mass at the rate $(\dot{m})$, then the temperature in the disc is given by

$$T(r) = \left(\frac{GM\dot{m}}{2\pi r^3 \sigma} \right)^{1/4}. \tag{4.1}$$

The temperature decreases outwards as the inverse three-quarter power of the radius and is proportional to the quarter power of the accretion rate.

Accretion onto stellar-mass objects

Figure 12 plots the temperature of the accretion disc around a solar-mass object that is accreting at a rate $\dot{m} = 10^{-8}$ $M_\odot$/ yr, which are typical of the values inferred for many binary stars, where one star is dropping mass onto its companion (Chapter 3, 'Binary Stars').

The horizontal scale, which like the vertical scale is logarithmic, extends from the radius of a solar-mass black hole, marked $R_s \simeq 3$ km, up to radii beyond the orbit of Pluto (page 72). The temperature falls from 100 million degrees at R_s, a factor of several hotter than the core of the Sun, through the temperature $T_\odot$ of the Sun's surface at about the solar radius, down to about 100 K at the radius of the Earth's orbit. If the accreting object is a black hole, essentially the whole radial range plotted is physically relevant, while if the accreting object is a solar-mass white dwarf, only the part to the right of the line marked R_{wd}, where $T = 200,000$ K, is physically significant, and if the accreting object is a star like the Sun, the only relevant region is that to the right of

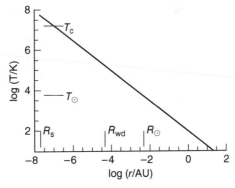

12. **Temperature at radius (r) in an accretion disc around a compact object of one solar mass. An Astronomical Unit (AU) is the mean radius of the Earth's orbit (page 8). The accretion rate is assumed to be $10^{-8} \, M_\odot \, yr^{-1}$. Also marked are the temperatures T_c and $T_\odot$ at the centre and surface of the Sun and the radii R_s, R_{wd}, and $R_\odot$ of a solar-mass black hole, a typical solar-mass white dwarf star, and the surface of the Sun.**

the line marked $R_\odot$, where $T = 4,700 \simeq T_\odot$. These temperatures are such that the disc will mostly radiate X-rays just outside R_s, soft X-rays, and ultraviolet light just outside R_{wd}, and mostly optical photons just outside $R_\odot$.

Figure 13 plots the luminosity radiated by the accretion disc from points outside the radius (r) on the horizontal axis. We see that the disc radiates only a few thousandths of a solar luminosity from beyond the Earth's orbit, about $L_\odot/2$ from beyond the radius of the Sun, about $60 \, L_\odot$ from beyond the radius of a white dwarf and $\sim 100,000 \, L_\odot$ down to the radius of a black hole. Taken with the conclusions we drew from Figure 12, it follows that when accreting at the rate $10^{-8} \, M_\odot \, yr^{-1}$, a solar-mass black hole will be an extremely luminous hard X-ray source, a white dwarf will be a luminous soft X-ray source, a main-sequence star will receive a significant boost to its luminosity from the accretion disc. In all these cases the portion of the disc that lies outside the Earth's orbit

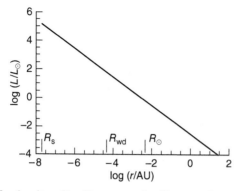

13. The luminosity radiated by an accretion disc around a solar-mass object outside radius r. Radii are given in units of the radius of the Earth's orbit and luminosities in solar luminosities. The accretion rate is assumed to be 10^{-8} $M_\odot$ yr^{-1}.

will contribute only infrared radiation and will be hard to detect against the luminosity from the inner portions of the disc and the accreting body. Nonetheless, this portion of the disc is of great importance as the site of planet formation (Chapter 5, 'Birth of planets').

Quasars

Figures 12 and 13 give results characteristic for stellar-mass black holes. A very different scaling is required for the black holes that sit at the centres of galaxies. These have masses in the range 10^6 to 10^{10} $M_\odot$ and when they are powering quasars they must have accretion rates of order 1 $M_\odot$ yr^{-1}. So these objects are $\sim 10^8$ times more massive that stellar-mass black holes and they accrete $\sim 10^8$ times faster. Since the characteristic radius R_s of a black hole is proportional to its mass, these holes are $\sim 10^8$ bigger in radius too. From equation (4.1) we see that at a given multiple of the black-hole radius R_s, the disc around a supermassive black hole has a temperature that is lower by a factor 100, and a luminosity which is a factor 10^8 greater than the corresponding values for a solar-mass black hole. Figures 14 and 15 reflect these facts.

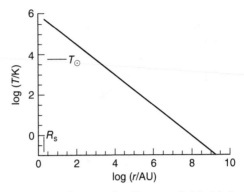

14. The temperature of an accretion disc around a black hole of mass $10^8 \, M_\odot$ like those found at the centres of galaxies when the accretion rate is $1 \, M_\odot / \text{yr}$. R_s marks the radius of the black hole. The Sun's surface temperature $T_\odot$ is also marked for reference.

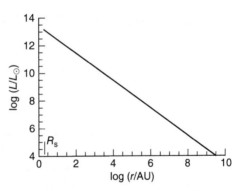

15. The luminosity radiated outside radius (r) for a disc around a $10^8 \, M_\odot$ black hole that is accreting at a rate $1 \, M_\odot / \text{yr}$.

The characteristic radius of the black hole is now slightly bigger than the Earth's orbit and the temperatures reached there are only $\sim 100{,}000$ K so the bulk of the radiation will emerge in soft X-rays and ultraviolet light. The luminosity of such a system is staggering. A luminosity equivalent to that of our entire Galaxy of about 100 billion stars is radiated from the portion of the disc beyond 600 AU, so a factor ten larger than the radius of Pluto's orbit. A hundred times more luminosity emerges between that radius and the surface of the black hole.

On account of this prodigious luminosity, much of it emitted in the readily observed ultraviolet and optical bands, accreting supermassive black holes can be observed right across the Universe. When spectra of quasars were first obtained, astronomers failed to interpret them correctly because they were unprepared for the large shifts to the red of their spectral lines. This shift is quantified by the *redshift z*, which is defined by the relationship between the wavelengths at which light is emitted by the quasar and observed on Earth:

$$\lambda_{obs} = (1 + z)\lambda_{emit}.$$

Hence $z = 0$ implies no redshift, while $z = 1$ indicates that spectral lines are observed at twice the wavelengths at which they are emitted. In 1963 Maarten Schmidt dropped an intellectual bombshell by demonstrating that the spectrum of the object known as 3C-273 has a redshift $z = 0.158$. In the wake of Schmidt's paper, spectra of many other sources were shown to have even larger values of z, and the conventional interpretation of these redshifts was that they were caused by cosmic expansion. However, no galaxy was then known to have such a large redshift and many astronomers were sceptical that these sources could be as distant and luminous as the cosmological interpretation of the redshift required. Tens of thousands of the *quasi- stellar objects (QSO)* are now known, and some have redshifts in excess of $z = 7$!

Journey's end

After spiralling through the accretion disc almost to the radius of the accreting body, gas has to pass from the disc to the body, and the way it makes this transition impacts on the structure of the accretion disc behind it. Consideration of the case of an accreting white dwarf will reveal what's at issue. At the inner edge of the accretion disc gas is effectively on a circular orbit that skims over the surface of the star. The kinetic energy that keeps it in orbit is $E_c = \frac{1}{2}GM/r$ per unit mass, which is equal in magnitude to all the energy the gas would have lost if it had spiralled from infinity to r. Hence it is a prodigious amount of energy, and if the gas crashes into the star, this energy will be turned into heat in a flash. Since gas does ultimately crash into the star, a thin layer of exceedingly hot gas develops at the interface between the star and the disc. This *boundary layer* naturally boosts the emission of the system at the shortest wavelengths.

Recall that three quarters of the energy radiated by each annulus in the body of the disc arises from the difference between the rate at which work is done on that annulus by the annulus interior to it, and the rate it works on the next annulus out. The innermost annulus is in an anomalous situation because it does work on both the annulus outside it and the boundary layer, which spins slowly because it is braked by the star. So while the boundary layer glows unusually brightly, the innermost annulus of the disc glows more faintly than usual.

The situation we just described, in which the accretion disc extends right in to a slowly rotating solid star is not universal. Some accreting bodies (especially neutron stars and white dwarfs) have powerful magnetic fields embedded in them. Magnetic field lines emerge from the star, loop through the space around the star and re-enter the star at another location, just as field lines emerge from the Earth in the Arctic and plunge back into the Earth in the Antarctic. If a field-carrying star spins, the field lines whip round the star, and if the magnetic field is sufficiently strong they can exert significant forces on gas even at large distances from the star where they are rotating at a significant fraction of the speed required for a circular orbit. In these circumstances at some critical radius r_B gas moves from the accretion disc onto a passing field line (which is moving slowly relative to the orbiting gas) and

16. Gas is lifted off a disc by DQ Her.

then runs along the field line towards a pole of the star (Figure 16). In this situation there is no thin, hot boundary layer, but rather two hot spots, one near each pole of the star, where gas streaming along field lines suddenly crashes into the star.

As gas streams along field lines towards the star, is transmits its angular momentum through the field to the star, thus increasing the star's rate of spin. As the spin rate increases, the radius r_B decreases and the disc extends further in. The logical endpoint of this progression is when r_B shrinks to the radius of the star and the star is spinning so fast that matter on its equator is effectively in orbit. The star is then said to be spinning at *breakup*. In practice this extreme point is never reached, but accretion discs do spin neutron stars up to a significant fraction of the breakup rate.

Discs around black holes end in yet another way. Einstein's theory of general relativity decrees that the circular orbit around a non-rotating black hole that has the largest angular velocity has a radius of $6R_s$. Consequently, already at this radius we have an annulus that does work on the annuli either side of it, so is dimmer than would be the case in Newtonian mechanics. At $3R_s$ circular orbits become unstable and particles can simply spiral into the black hole without losing any energy: their energy is swallowed by the black hole. Hence there is no hot boundary layer around a black hole, and the innermost part of the accretion disc is fainter than our Newtonian calculations predict. The black hole is spun up in the same way that a white dwarf is.

Impact of the magnetic field

In Chapter 3, 'Stellar coronae', we saw that magnetic field lines are dragged along by conducting fluid, and are constantly exchanging energy with the fluid because a field line generates tension that pulls on the fluid. Because an accretion disc is in differential rotation, two initially neighbouring points soon diverge from one another unless they lie exactly the same distance from the disc's rotation axis (Figure 17). If these two points lie on the same field line, it follows that the field line is stretched, and the tension it generates increases. Now from Figure 17 we see that as the field line is stretched, the tension opposes the rotation of the mass element that's initially at a smaller radius, and pulls the outer mass element in the direction of rotation. The field will, in fact,

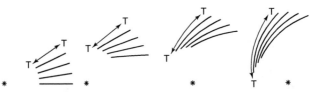

17. Four snapshots of magnetic field lines in an accretion disc being stretched by the disc's differential rotation. In each snapshot the star is shown at the bottom, and the earliest snapshot is on the extreme left. The disc rotates counter-clockwise. In that snapshot the field lines are short and run radially. By the final snapshot on the extreme right, the field lines are longer and are becoming tangential.

transfer angular momentum from the inner mass element to the outer mass element. But this is just what we expect viscosity to do! Moreover, by stretching the field line the differential rotation will strengthen the magnetic field, so no matter how weak the field was initially, it will amplify until it is strong enough to modify the dynamics sufficiently to limit the endless stretching of field lines. We have just described the essential physics of the *magnetorotatonal instability (MRI)*, which in 1991 was proposed by Steve Balbus and John Hawley as the origin of viscosity in accretion discs.

Jets

The advent of the MRI not only solved the longstanding puzzle of why the viscosity within accretion discs was as large as the observations required, but more significantly it offered a clue as to the origin of the most remarkable aspect of accretion discs: jets. Wherever we have reason to believe a compact object is accreting, the observational data are dominated by *outflow* rather than inflow. A Herbig-Haro object such as that shown in Figure 18 is the classic example: the object's core is a forming star, but its most conspicuous feature is a pair of narrow jets along which very cold gas is racing at $\sim 200\,\mathrm{km\,s^{-1}}$, way faster than the sound speed in the gas. On Earth aerospace engineers can only dream of replicating this trick.

Another remarkable example of jet formation is offered by the object SS433, which consists of a star rather more massive than the Sun in orbit around a black hole. The ordinary star is feeding gas to the black hole, and the gas spirals onto the black hole through an accretion disc. Two jets carry material along the spin axis of the accretion disc at over a quarter of the speed of light ($0.26c$, where $c = 300{,}000\,\mathrm{km\,s}$ is the speed of light), yet the gas is cool enough to contain hydrogen atoms and emit the characteristic spectral lines of hydrogen. So again the gas in the jet is moving much faster than the sound speed.

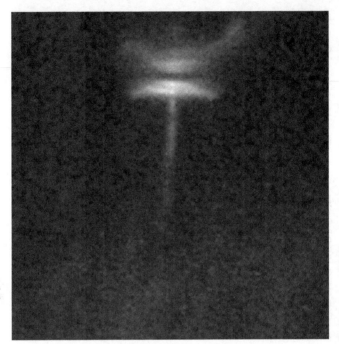

18. The Herbig-Haro object HH30. Jets emerge either side from a disc accreting onto a proto-star. The disc itself is dark and flared; we see it as a silhouette against a bright background of scattered light.

In SS433 the spin axis of the accretion disc precesses around a line that lies within 11° of the plane of the sky (Figure 19), and at their bases the jets travel along the direction of the current spin axis. As each parcel of gas moves away from the black hole, it travels in a straight line but behind it the disc is precessing and launching fresh parcels along its new spin axis. So overall the jets spiral round a cone that has an opening angle of 40° (Figure 20).

Our final example of a jet is on an altogether grander scale: the radio galaxy Cygnus A shown in Figure 21. Two extremely thin jets

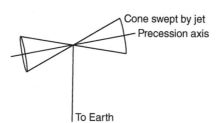

19. The geometry of SS433. Oppositely directed jets sweep over the surface of a cone. This cone has an opening semi-angle of 20° and its axis is inclined at 80° to the line of sight to Earth.

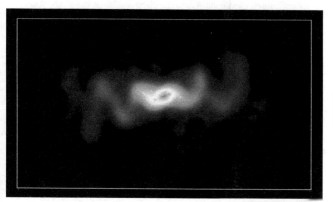

20. A radio image of the corkscrew jets of the binary star SS433.

of plasma emerge from the centre of a giant elliptical galaxy and 0.77 Mpc from the black hole slam into the intergalactic medium with incredible force, in the process accelerating electrons to stupendous energies (Chapter 6, 'Shocks and particle acceleration'). These energetic electrons emit the radio waves with which the system is imaged in Figure 21.

These three examples differ in many respects. The first involves accretion onto a young star while the last two involve accretion

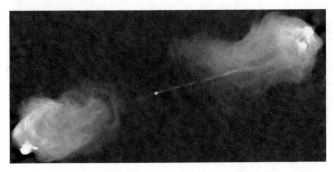

21. A radio image of the radio galaxy Cygnus A. Thin jets emerge from the galactic nucleus and slam into the circumgalactic gas more than 60 kpc away.

onto black holes. The first two are driven by stellar-mass objects while the last is driven by an object $\sim 10^8$ times more massive. The first involves completely non-relativistic jets, the second mildly relativistic jets and the third involves distinctly relativistic jets in the sense that particles in the jets have kinetic energies that exceed by a factor of a few their rest-mass energies (Chapter 6, 'Rest-mass energy'). Despite these great differences of scale each system consists of a pair of narrow jets which are carrying material away from the accreting object extremely supersonically. The ability of Nature to build systems that differ enormously in length and velocity scale but can nevertheless be roughly rescaled into each other implies that the physics of these systems is simple in the sense that it is restricted to electromagnetism and gravity, two theories that lack natural scales. By contrast, on Earth nearly all phenomena involve quantum mechanics, which introduces a scale through Planck's constant $h = 6.6 \times 10^{-34}$ J s. Stars, planets, the interstellar medium all have scales imposed on them through quantum mechanics and h. Jets seem to be a rare phenomenon that is independent of h. Since they are simple, you might imagine we understand how they form. Sadly, this is far from the case.

Driving jets

Although our understanding of jet formation is incomplete, we think we understand the basic principles. We have seen that we expect accretion discs to be laced by constantly amplified magnetic field lines. In Chapter 3, 'Stellar coronae', we saw that the Sun's surface is also laced by constantly stretched and twisted magnetic field lines, and that release of magnetic energy by reconnecting field lines in the low-density plasma above the photosphere drives a wind of plasma away from the Sun, past the Earth, and off into interstellar space. Something very similar must happen just above and below the mid-plane of an accretion disc, so the space above and below the disc is filled with gas that is too hot to be confined by the system's gravitational field and therefore flows away from the system. But in this case, unlike that of the Sun, the flow is somehow collimated into a narrow jet. It is likely that on account of the systematic rotation of the disc, the outflowing gas is confined by a helices of field lines that twist around the outflowing gas rather as a boa constrictor wraps itself around its prey to crush it to death. The tension in these field lines restricts the expansion of the ultra-hot gas in directions perpendicular to the disc's spin axis, so the gas instead expands in the direction of the spin axis, speeding up, and cooling down, as it does so. In this way a narrow column forms of gas that's moving much faster than its sound speed.

The process we have just described must happen at many radii simultaneously—this is a consequence of the scale-free nature of disc dynamics. At large radii the disc is cooler and rotating more slowly than at small radii, so we expect gas accelerated from these radii to achieve a lower ultimate flow velocity than gas accelerated from small radii. Hence we expect a jet to be a nested structure with a fast core surrounded by cylinders in which the flow velocity decreases steadily outwards. Even though the flow in the disc is never more than mildly relativistic (Chapter 6), ultra-relativistic jets are produced by accretion onto neutron stars as well as black holes. We do not understand how Nature achieves this feat.

High-efficiency jets

The emergence of nested jets from each side of an accretion disc requires a profound modification of the model of an accretion disc introduced in the section, 'Basic disc dynamics' at the start of the chapter. For that model was based on the assumption that the rate of flow of matter through the disc is the same at all radii. If gas leaves the disc at each radius to flow out in the jet, the rate of flow of matter through the disc must decrease as we move inwards. Moreover, previously, viscosity had to carry outwards angular momentum at the same rate that the flow of matter was carrying angular momentum in. When there is a jet, viscosity is not the only mechanism that takes away the angular momentum that is carried past radius r by the flow of matter: the jet carries angular momentum away from every point interior to r, so the viscous flow of angular momentum at r is smaller than it would be in the absence of a jet. Since this flow of angular momentum is responsible for three quarters of the heat input at r, the introduction of jets causes the disc to cool and to radiate less strongly. In fact, the energy output of the disc is being shifted from heat (in the form of radiation) to mechanical energy (in form of the jet's kinetic energy). Observations indicate that this shift can be almost complete, so almost all the energy released by the accretion disc emerging as kinetic energy.

Time-domain astronomy

We have studied accretion discs that have reached a steady state. However, the luminosities of real accretion discs tend to vary quite a lot, and a great deal can be learnt about a disc and the system in which it is located by monitoring the system's *light curve*, the luminosity as a function of time. More information still can be extracted if spectra taken at different times are available.

In some systems a significant quantity of gas is regularly dumped on a small region of an accretion disc. The matter is then quickly spread by differential rotation into an annulus of enhanced density. Then on a longer timescale viscosity carries angular momentum from the inner edge of the annulus to its outer edge, with the result that the inner edge moves inwards and the outer edge moves outwards. Hence the initially thin annulus of enhanced density becomes an ever broader region. Just as different radii within a steady-state disc heat to different temperatures, so the temperature of the inner edge of the annulus rises, and that of the outer edge falls, so the spectrum of radiation emitted by the whole annulus evolves away from a near black-body spectrum towards the kind of spectrum emitted by a steady-state disc. The timescale on which this evolution occurs is inversely proportional to the magnitude of the viscosity, which we did not need to know to derive the properties of a steady-state disc.

In some systems surges in luminosity are caused by a parcel of gas being deposited on the accretion disc as described above, but in others surges are caused by sudden changes in the magnitude of the viscosity. When the viscosity is low, matter takes a long time to move through the disc, so when the disc is in a steady state a given accretion rate onto the star corresponds to a large density of gas in the accretion disc. Conversely, when the viscosity is large the steady-state density in the disc is low but the luminosity is the same. Hence a sudden increase in viscosity within a disc that has reached a steady state causes matter to drain out of the disc onto the star faster than it is dropping onto the disc, and the luminosity surges before relaxing back to its original value as the high-viscosity steady state is approached.

What causes the viscosity to switch between high and low values? This is not well understood, but the mechanism is probably connected to the fact that the viscosity is generated by turbulence in the disc, and the turbulence is itself powered by viscosity. So in the low-viscosity state the turbulent eddies are small and the viscosity generated by these eddies is small, while in the high-viscosity state large eddies generate a large viscosity.

Many low-mass *X-ray binaries* regularly transition between a state in which the system has a soft X-ray spectrum and one in which it has a hard spectrum and a lower luminosity—the system, which contains an accreting black hole or neutron star, is alternating between low-hard and high-soft states. Often, but not always, relativistic jets squirt out as the system switches from its high-soft to its low-hard state. It is thought that this transition occurs when gas near the centre of the accretion disc is ejected along the spin axis, leaving a low-density region around the accreting body. In the high-soft state the dense central region radiates like a black body, while in the low-hard state the gas is too tenuous to create many photons itself. Instead it contributes to the system's luminosity by the *inverse Compton process* in which a relativistic electron collides with a photon. Just as a footballer's boot energizes the ball at a free kick, the electron can greatly increase the energy of the photon. In this way an infrared photon can become an X-ray or even gamma-ray photon and the spectrum of the object is hardened. X-ray binaries that sometimes fire jets are called *micro-quasars*.

The jets cause a micro-quasar's radio-frequency luminosity to increase by a factor of order 100—we'll discuss the physics of jets in Chapter 6, 'Rest-mass energy'. Consequently, at radio frequencies these sources are more variable than in X-rays, and when they are *radio loud* their radio emission is dominated by jets.

We have seen that the black holes that power quasars have radii R_s that are $\sim 10^8$ times larger than the radii of the stellar mass black holes that drive some X-ray binaries. The characteristic velocity of both types of systems is the same—a good fraction of the speed of light—so we expect quasars to vary on timescales that are longer by a factor $\sim 10^8$ than the timescales on which X-ray binaries vary. For example, a second in the life of an X-ray binary is equivalent to three years for a quasar, and the year or so between the changes of state on an X-ray binary is equivalent to 100 Myr for a quasar. Hence during our puny lifetimes we cannot expect to observe changes of state by quasars, but we do expect to find the population divided into radio quiet and radio loud populations. In fact before the connection between quasars and micro-quasars was recognized quasars had been divided into radio-quiet and radio-loud quasars, with more than 90 per cent of quasars being radio quiet. This proportion is in line with the fraction of their time during which micro-quasars are radio quiet.

At the blue end of the electromagnetic spectrum (X-rays or ultraviolet radiation depending on the nature of the accreting body) the brightness of many accreting systems flickers at a characteristic frequency—one speaks of *quasiperiodic variability*. Because quasiperiodic variability is concentrated at the blue end of the spectrum, which is dominated by emission from the inner edge of the disc, its characteristic frequency is thought to be the orbital frequency at the inner edge of the disc. Hence it tells us about the nature of the accreting body.

In micro-quasars quasiperiodic variability has a characteristic timescale of a millisecond, so in quasars the equivalent timescale is a day. The amplitude of these fastest fluctuations is small. On a timescale that's longer by a factor of a few hundred the X-ray luminosity of a micro-quasar can change by a good fraction of itself, and in quasars similar fluctuations happen on the timescale of a year. These fluctuations are of considerable diagnostic value.

For example, they can be used to estimate the mass of a quasar's black hole. When the luminosity of the accretion disc increases, gas in clouds that orbit the black hole at some distance is stimulated by ionizing photons from the inner disc to strengthen its optical and UV emission line emission. But there is a delay T between the luminosity of the accretion disc rising and the emission lines strengthening because the ionizing radiation takes time to cover the distance $r = cT$ from the quasar to the orbiting gas. The orbital speed v of the emitting gas can be estimated from the width of the emission lines, so using the formula $v^2 = GM/r$ for the speed of a circular orbit, we find that the mass of the black hole is cTv^2/G.

In conjunction with strong gravitational lensing (Chapter 6, 'Gravitational lensing'), fluctuations in the luminosities of quasar accretion discs can also be used to determine the scale of the Universe and to search for lumps of dark matter.

Chapter 5
Planetary systems

Astrophysics started with Newton's work on the dynamics of the solar system (Table 1), and work to understand how the solar system formed and evolved to its present state is at the forefront of astrophysics to this day.

In 1995 Michel Mayor and Didier Queloz announced the discovery of a planet orbiting the star 51 Pegasi, which is not unlike the Sun. Since then roughly a thousand extra-solar planetary systems have been discovered, and the process of understanding how these systems formed and evolved to their present states is having a profound impact on how we think about our own planetary system, and indeed our place in the Universe. Our understanding of the formation and evolution of planetary systems is developing rapidly, but we still don't know how unusual our system is.

Dynamics of planetary systems

Newton showed that if planets moved in the gravitational field of the Sun alone, their orbits would be ellipses (Figure 22). He was aware that this demonstration was only the start of a long journey towards understanding the complete dynamics of the solar system, because planets have non-zero masses and one needs to consider the impact of the mutual gravitational attraction of the planets. This undertaking was at the forefront of mathematical physics for

Table 1. The solar system The symbol $\mathcal{M}_\oplus$ denotes Earth's mass, while P_J is the period of Jupiter. We include Pluto although in 2006 the International Astronomical Union deprived it of the dignity of being called a planet on the grounds that it is merely a large Kuiper-Belt object.

Planet	$M/\mathcal{M}_\oplus$	a/AU	e	i	Period	$P:P_J$
Mercury	0.055	0.387	0.206	6.34	0.241	1:49.2
Venus	0.815	0.723	0.007	2.19	0.615	1:19.3
Earth	1	1	0.017	1.58	1	1:11.9
Mars	0.108	1.524	0.093	1.67	1.881	1:6.31
Jupiter	317.8	5.203	0.049	0.32	11.86	
Saturn	95.15	9.582	0.056	0.93	29.46	2.48:1
Uranus	14.54	19.19	0.047	1.02	84.02	7.08:1
Neptune	17.15	30.07	0.009	0.72	164.8	13.9:1
Pluto	0.002	39.26	0.245	17.1	247.7	20.9:1

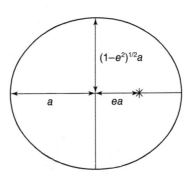

22. An orbital ellipse of eccentricity $e = 0.5$. The principal axes have lengths a and $\sqrt{1-e^2}\,a$. The centre of attraction, marked by a star, lies distance ea from the centre of the ellipse.

the following 250 years. It culminated in the work of Urbain Le Verrier (1811–77), who showed that Newtonian physics left a very slight anomaly in the orbit of Mercury. In 1916 an argument for the correctness of Einstein's brand new theory of general relativity was that it accounted for this anomaly very naturally.

In the last twenty years two developments have revived interest in planetary dynamics. The first was the availability of fast and relatively cheap computers, which made it possible for the first time to integrate the full equations of motion for billions of years, and the second was the discovery of extra-solar planetary systems, which are often dramatically different from ours and got astrophysicists wondering why that is so, and what it has to tell us about the solar system.

Disturbed planets

Since the masses of the planets are much smaller than that of the Sun, the natural approach to planetary dynamics is *perturbation theory*: we put the planets on the orbits they would have if they were all massless, and ask how a given planet's orbit will evolve in response to the force on it from the other planets. In this scheme the orbit of a planet is at all times one of Newton's elliptical orbits around the Sun, but the orbit in question slowly changes in response to perturbations from other planets.

The key numbers quantifying an orbital ellipse are its *semi-major axis a*, which controls the orbit's energy ($E = -GM/2a$), its *eccentricity e*, which describes the shape of the ellipse (Figure 22), and its *inclination i*, which is the angle between the plane of the ellipse and the *invariable plane*, an imaginary plane that is defined by the solar system's angular momentum (Figure 23). Table 1 lists these quantities for the Sun's planets. We use perturbation theory to compute how these numbers change over time.

Angular momentum is a crucial quantity in planetary dynamics, as in the dynamics of accretion discs (Chapter 4, 'Basic disc

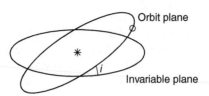

23. The inclination _i_ is the angle between the invariable plane and the plane of the planet's orbit.

dynamics'). For a fixed semi-major axis a, the angular momentum

$$L = \sqrt{1 - e^2}\sqrt{GMa} \tag{5.1}$$

is largest when the eccentricity $e = 0$ and the orbit is circular, and decreases to 0 as e tends to unity.

We imagine that the mass of each planet is spread out along its orbit, so each orbit becomes an elliptical wire of slightly non-uniform density—the wire is densest where it is furthest from the Sun because at this point the planet moves most slowly (Figure 24). These wires attract each other gravitationally, and by virtue of the fact that they are elliptical and do not lie in the same plane, they exert torques (page 51) on one another as indicated in Figure 24. Torque gives the rate of change of angular momentum (Chapter 4, 'Basic disc dynamics'), so planets exchange angular momentum. To the extent that it is a valid approximation to replace planets by elliptical wires, planets do not exchange energy, so each planet's semi-major axis a is fixed while its eccentricity e changes.

If the planets had negligible mass, the orientation of the long axis of each planet's ellipse would remain fixed in space. If the effect of the mass of the planetary system were the same as that of a thin axisymmetric disc of matter in the invariable plane, the long axes of the planetary ellipses would rotate slowly in the sense opposite to that in which in which planets rotate around their ellipses. This backwards motion of the long axes is called _precession_.

74

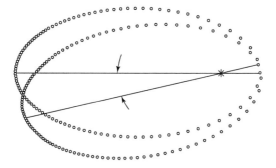

24. The positions of two planets on similar orbits are shown at a hundred equally spaced times. We represent each planet by an elliptical wire with mass density proportional to density of points drawn here. The arrows show the direction of the torque experienced by each ellipse due to the gravitational pull of the other ellipse.

In this axisymmetric model of the planetary system, each planet has its own precession frequency. Consequently, the torque that one planet applies to another keeps changing sign because over time the angle between the long axes of two planet's ellipses is as often such that planet 1 transfers angular momentum to planet 2 as it is such that the flow of angular momentum is from 2 to 1. In these circumstances the eccentricity of each planet's orbit oscillates slightly but nothing more interesting occurs. In particular, the angle between the long axes of two planets' ellipses constantly increases.

It can happen that the precession frequencies Ω_1 and Ω_2 of two planets are nearly *resonant* in the sense that

$$n_1\Omega_1 \simeq n_2\Omega_2, \tag{5.2}$$

where n_1 and n_2 are small integers. Then planet 1 can transfer angular momentum to planet 2 for a time long enough that the eccentricities of both planets change significantly. As a planet's eccentricity changes, so does the its precession rate, so the rate at which the angle between the long axes of the ellipses increases becomes non-uniform. In fact this angle can oscillate instead of continually increasing. We say that the angle between the ellipses is *librating* rather than *circulating* as it does when no resonant condition (5.2) holds.

Resonances like that just described are the key to understanding many phenomena in any branch of physics where small disturbances are involved because a small disturbance can be important only if it acts in the same way for a long time. In the absence of a resonance, the sense of the disturbance is constantly changing, so its time-averaged effect is zero; a resonance gives a weak disturbance the opportunity to act in the same sense for a long period, and thus to effect significant change.

Astrophysics

A resonance that arises when we replace the planets by elliptical wires is called a *secular resonance* to distinguish it from a more fundamental resonance between the frequencies Ω_ϕ at which the planets move around their ellipses—in the wire model this motion has been averaged away. Two planets are in a *mean-motion resonance* when there are small integers n_1 and n_2 such that

$$n_1 \Omega_{\phi 1} = n_2 \Omega_{\phi 2}. \qquad (5.3)$$

Planets that are in a mean-motion resonance can exchange energy as well as angular momentum. Hence their semi-major axes a can change in addition to their eccentricities.

Birth of planets

A very young star is always surrounded by a disc from which it accretes matter. As the mass of the star increases, the temperature rises in its core (Chapter 3, 'Star formation'), and if the star's mass exceeds $0.08\,\mathrm{M_\odot}$, nuclear burning starts up there (Chapter 3, 'Nuclear fusion'). As the luminosity of the star grows, its radiation heats the surrounding disc, and the warmed gas tends to escape from the gravitational field of the young star back into interstellar space. Particles of dust in the disc are too massive to escape, so the ratio of dust to gas in the disc increases as the young star warms its disc. In the increasingly dusty disc, dust particles collide and merge to form bigger particles. Eventually the most massive dust particles are kilometres in diameter and their gravitational fields are strong enough to deflect significantly the velocities of nearby gas and dust. An *asteroid* has formed.

Gradually asteroids collide and merge to form bigger and bigger asteroids. The self-gravity of the most massive asteroids pulls them into nice spherical balls, which become radially structured as dense material sinks and less dense material rises. These balls are planetary cores.

If a very massive core forms sufficiently early on, before all the gas has been dissipated at its radius, the core may trap some of the gas in its gravitational field. This is how the massive outer planets Jupiter, Saturn, Uranus, and Neptune formed, while the rocky inner planets Mercury, Venus, Earth, and Mars failed to acquire significant quantities of gas because their gravitational fields are not strong enough to retain hydrogen and helium at the relatively warm temperatures prevailing in the inner solar system.

Evolution of planetary systems

Early on, when no dust particle has a dynamically significant gravitational field, everything, gas and dust, is on nearly circular orbits in the system's invariable plane. Later the gravitational field of each young planet drives a spiral wave through the disc (Figure 25). The planet is gravitationally attracted to the nearby parts of the spiral overdensity. The inner region pulls it in the direction of rotation so it acquires angular momentum from this region, while it is pulled back by and gives angular momentum to the outer region. A detailed calculation shows that it loses more angular momentum than it gains. If the planet is quite massive,

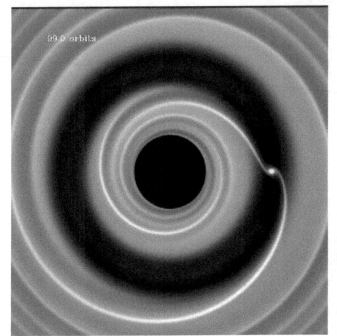

25. A planet (right centre) orbiting an (invisible) central star excites a spiral wave in the surrounding gas disc. Through this wave the planet gets angular momentum from the disc inside its orbit, and looses angular momentum to the exterior portion of the disc. These angular-momentum transfers create a region of near zero density around the planet.

these angular-momentum exchanges cause an annulus of low density to form around the planet. Material that was in this *evacuated annulus* has been either swept into the planet or pushed to beyond the edge of the annulus.

When an orbiting body loses angular momentum, it moves inwards. So the planet moves slowly towards the star. The material beyond the inner edge of the evacuated annulus would move in too as a result of surrendering angular momentum to the asteroid, but so long as plenty of gas is present, viscosity transfers angular momentum out through the disc fast enough to replace the angular momentum it has lost to the planet. Similarly, beyond the outer edge of the evacuated annulus, viscosity carries away the angular momentum provided by the asteroid and prevents the outer edge moving out. So the planet and its evacuated annulus moves slowly inwards *shepherding* the material of the disc so it remains in good gravitational contact with the planet at a safe distance on both sides of the evacuated annulus.

The rates at which planets drift inwards are not all the same, so the inner edge of the evacuated annulus of one planet can come into contact with the outer edge of the evacuated annulus of another planet. Numerical simulations of gas discs with embedded planets indicate that two planets are then likely to fall into a mean-motion resonance (see equation (5.3)) and become able to exchange both energy and angular momentum. It turns out that these exchanges lock the planets into the mean-motion resonance. Thus the inner planet picks up energy and angular momentum from the dust and gas beyond the inner edge of its evacuated annulus, while the outer planet gives energy and angular momentum to the dust and gas beyond the outer edge of its evacuated annulus. The inner planet gives the outer planet the amounts of energy and angular momentum required for them to stay in mean-motion resonance. When a planet trades energy and angular momentum with the disc on its own, it always loses out and ends up drifting inwards. But when two planets work in partnership, they aren't necessarily losers, and they may move very slowly outwards, or very slowly inwards.

The young solar system

It is thought that the young Saturn bumped into the young Jupiter in the way we have described and entered a 2 : 3 mean-motion resonance with Jupiter (three Jupiter years taking as long as two Saturn years) and then the two planets working in partnership pretty much stopped drifting inwards. Then the next planet out, we'll call it ice giant 1, drifting inwards encountered the almost stationary Jupiter–Saturn pair and entered a mean-motion resonance, probably again 3 : 2 with Saturn, and the system of three locked planets drifted only very slowly in radius. So along comes the next planet out, ice giant 2, and enters a mean-motion resonance with ice giant 1. This resonance may have been 3 : 4. Now all four planets, working in partnership, remained in pretty much the same places while the young Sun dispersed the gas in the disc.

Since the orbital time around a star increases with radius r as $r^{3/2}$, the time taken for dust to accumulate into asteroids and asteroids to gather into planets increases as we move outwards, and beyond ~ 20 AU this process was still incomplete when the Sun had dispersed the gas. So there was no ice giant 3 beyond ice giant 2, only a large number $\sim 1,000$ of Pluto-sized objects and zillions of asteroids. Once the gas was gone, there was nothing to damp the eccentricities of the asteroids that were excited by the gravitational fields of the Pluto-sized bodies. This ensemble of asteroids and Plutos surrounded the four locked planets but did not extend into the evacuated annulus of ice giant 2. The present-day Kuiper Belt of asteroids is the descendent of this ensemble and we shall refer to the ensemble as the *Primordial Kuiper Belt (PKB)*.

One of the ice giants, probably ice giant 1, was on a slightly eccentric orbit and was able to exchange energy and angular momentum with the asteroids near the inner edge of the PKB at $r < 20$ AU. If it wasn't locked in mean-motion resonances to the other planets, it would respond to this loss by moving to a fairly circular orbit of smaller radius. But it is locked, so it responds by moving to a more eccentric orbit. For a while its eccentricity steadily grows and then suddenly a secular resonance within the four-planet system causes the eccentricity of ice giant 1 to decrease, and the angular momenta of the planets to change in such a way that the mean-motion resonant conditions are broken.

With the resonant conditions broken the planets can no longer exchange energy so any loss of angular momentum will lead to an increase in eccentricity (see equation (5.1)). The eccentricities of the two ice giants quickly grow to large values so each of these planets crosses the other's orbit and possibly even Saturn's orbit. This is a time of great peril for the solar system, for a planet on a highly eccentric orbit is likely to induce other planets to move to eccentric orbits, and once Jupiter was on a highly eccentric orbit it would not be long before Jupiter would have driven every other planet either into the Sun or completely out of the solar system.

We shall see below that a catastrophe of this type has probably occurred in many planetary systems. We owe our existence to good fortune and the way the PKB acted as a fire bucket. As the eccentricities of the ice giants grew, they penetrated into the PKB and started to have close encounters with asteroids and Plutos. Scattering objects in the PKB damped the eccentricity of the ice giants, and the system settled to its present configuration. Neptune is now in a 1 : 2 mean-motion resonance with Uranus and on an orbit of low eccentricity and semi-major axis $a = 30.1$ AU that places it far into the PKB. Even the orbit of Uranus probably lies within the PKB (Table 1). In some simulations of the evolution of the four-planet system after the resonance condition is broken, ice giant 1 ends up on a smaller orbit than ice giant two, and in other simulations it ends on the larger orbit. Thus we do not know which ice giant Neptune is.

The population of the PKB was decimated when the ice giants swept through it, so the present Kuiper belt contains only $\sim 0.07 \, \mathcal{M}_\oplus$ rather than the $\sim 40 \, \mathcal{M}_\oplus$ from which we believe it started, and all but one of the ~ 1000 Plutos and many of the asteroids have been turfed out of the solar system. However, many of these objects at some stage appeared within the PKB and were scattered by the planets from Mercury to Saturn, pitting their surfaces and damping their eccentricities. Indeed, the rate at which asteroids hit the moon can be determined from the pattern of craters they made, and long before our current picture of the evolution of the solar system emerged it was known that there was a *late heavy bombardment* (*LHB*) of the Moon approximately 0.7 Gyr after the formation of the Sun 4.6 Gyr ago. Another likely legacy of this period of high asteroid density are the *Trojan asteroids* of Jupiter, which move on the same orbit as Jupiter but on the other side of the Sun. It is thought that Jupiter captured these asteroids at this time.

Order, chaos, and disaster

Newton bequeathed us a wonderful tool for predicting the future: the differential equation (page 2). Use the numbers that describe the current configuration of a dynamical system as the initial conditions from which you solve the equation, and from the solution you can read off a prediction for the system's configuration at any time. Unfortunately in the 20th century it emerged that this scheme often doesn't work. The problem is that the behaviour of the solution can be *extremely* sensitive to the initial conditions. Henri Poincaré, a cousin of the man who led France in much of 1914–18 war, first sighted this problem at the beginning of the 20th century, but the scale of the problem only emerged in the 1960s as electronic computers made it feasible to solve the differential equations of generic dynamical systems. The solar system provides at once the cleanest and the most awe-inspiring example of this phenomenon.

The differential equations that govern the motion of the planets are easily written down, and astronomical observations furnish the initial conditions to great precision. But with this precision we can predict the configuration of the planets only up to ~ 40 Myr into the future—if the initial conditions are varied within the observational uncertainties, the predictions for 50 or 60 Myr later differ quite significantly. If you want to obtain predictions for 60 Myr that are comparable in precision to those we have for

40 Myr in the future, you require initial conditions that are 100 times more precise: for example, you require the current positions of the planets to within an error of 15 m. If you want comparable predictions 60.15 Myr in the future, you have to know the current positions to within 15 mm. We really are up against a hard limit to our knowledge here in the sense that it is inconceivable that we will be able to specify the current configuration of the solar system to be able to predict the positions of the planets more than ~ 60 Myr in the future. This is a disappointingly small fraction of the 4, 600 Myr age of the system.

In these circumstances all we can do is use the differential equations to compute which configurations are most likely in the future. We do this by randomly sampling current configurations that are consistent with the observational data, and for each sampled configuration computing the corresponding solution so we can predict where the planets will be in that case. The most probable future configurations are those that are nearly reached from many sampled current configurations.

An important feature of the solutions to the differential equations of the solar system is that after some variable, say the eccentricity of Mercury's orbit, has fluctuated in a narrow range for millions of years, it will suddenly shift to a completely different range. This behaviour reflects the importance of resonances for the dynamics of the system: at some moment a resonant condition becomes satisfied and the flow of energy within the system changes because a small disturbance can accumulate over thousands or millions of cycles into a large effect. If we start the integrations from a configuration that differs ever so little from the previous configuration, the resonant condition will fail to be satisfied, or be satisfied much earlier or later, and the solutions will look quite different.

The importance of resonances leads to astonishing sensitivity to the small print of physical law. In Chapter 6 we will introduce

Einstein's theory of relativity and discuss some of its astrophysical applications. Here we anticipate the fact that the importance of relativity is quantified by v^2/c^2, where v is a typical velocity and c is the speed of light. For the Earth this ratio is $\sim 10^{-8}$, so extremely small, and even for Mercury it's less than 2.5 times bigger. Yet an experiment conducted by Jacques Lasker and Mickael Gastineau indicates that we probably wouldn't be here if it weren't for the tiny effect that Einstein's correction to Newton has on the solar system.

Laskar and Gastineau evolved the solar system forward from its present configuration with two ensembles of solutions. In each case they used as initial conditions for the solutions configurations for the solar system that are consistent with the best observational data. They computed one set of solutions 5 Gyr into the future using Newton's theory without Einstein's tiny corrections, and the other set using Einstein's corrections. They found that Mercury attained an eccentricity $e > 0.7$ in just 1 per cent of the solutions when Einstein's corrections were included, whereas without Einstein's corrections only 1 solution out of the 2,500 computed kept Mercury on an orbit with $e < 0.7$. If Mercury does attain $e > 0.7$, the consequences for the Earth are dire, because an eccentric Mercury soon excites high eccentricity in Venus, which drives the Earth to high eccentricity, which drives Mars to high eccentricity. Various dramatic consequences ensue: in the 2,500 solutions to Newton's uncorrected equations, Mercury collided with Venus in eighty-six cases, and collided with the Earth in thirty-four cases. Even if the inner planets don't collide with one another, they are either driven into the Sun or flung right out of the solar system.

So it seems we live on the edge of a precipice, and were it not for Einstein's absolutely tiny corrections to Newton's equations, planet Earth would almost certainly not be in a position to offer us shelter. Even with Einstein's corrections, our security is not guaranteed longer than ~ 80 Myr into the future. Nevertheless, thanks to general relativity planet Earth is overwhelmingly likely to survive until it is engulfed by the swelling Sun $\sim 4\,$Gyr from now.

Einstein's corrections make life possible by de-tuning a resonance between Jupiter and Mercury. Because this is a weak resonance, it can have an impact only if conditions are just right. Einstein's corrections make it hard for them to be right.

Extra-solar systems

The first unambiguous detection of extra-solar planetary system was made a recently as 1995 by Michel Mayor and Didier Queloz, but now over a thousand planetary systems are known. We can gain insight into the evolution of planetary systems by considering the statistics of the configurations of the known systems: it's as if the Universe has computed a large number of solutions to evolutionary equations that are guaranteed absolutely correct.

We have seen that the solar system has retained eight planets on nearly circular orbits in the teeth of at least two grave threats: first about 700 Myr into its life when the tight formation of the giant planets was destabilized, and even today the stability of the inner solar system would be imperilled by changing its physics by a part in 100 million.

So it is not surprising that the first systems to be discovered resemble the endpoint of one of these catastrophes in that they consist of a Jupiter-like planet on a fairly short-period, eccentric orbit. However, the significance of this finding must be tempered by the fact that the observational technique used to find the first systems strongly favoured finding systems with a large planet on a short-period orbit: the systems were found by monitoring the velocities of stars for the periodic changes in velocity that signal motion of the star around the centre of mass of the star and its planet. The more massive and close the planet, the larger these velocity changes are and therefore the greater the likelihood that they will be detected above the noise.

Subsequently, a completely different technique has been used to detect planets, and with this technique systems with several planets on nearly circular orbits can be, and have been, found. The technique involves monitoring the brightnesses of stars to detect the slight drop in brightness as a planet passes between its star and Earth. The only systems that can be detected with this technique are those we view almost exactly edge-on, and data of the required precision can only be gathered from space. In May 2009, NASA launched the Kepler satellite to monitor stars in a small area of the sky. In the following four years Kepler definitely detected nearly a thousand planetary systems and drew up a list of several thousand stars that showed signs of planetary systems. Assessing the significance of these data is a very active field of research.

Chapter 6
Relativistic astrophysics

The familiar world of Newtonian physics is an approximation to relativistic physics, which is convenient and works well when relative velocities are significantly smaller than the velocity of light. Astronomers have identified many kinds of object that violate this restriction, and we have to use the full theory of relativity to model these objects.

We will outline the main results of relativity theory later, in 'Special relativity', but we need the most famous result now: $E = mc^2$. This equation summarizes the requirement that according to relativity, a particle's energy does not consist only of its kinetic and potential energies, as in Newtonian physics, but includes in addition its *rest-mass energy* $E_0 = m_0 c^2$, where m_0 is the mass the particle has when it is at rest. It is a number that is characteristic of the particle and never changes. By contrast the mass $m = E/c^2$ changes as work is done on or by the particle: when an electron is accelerated in a collider such as the Stanford Linear Collider (SLAC), its mass increases by a factor ~ 200. The ratio $\gamma = m/m_0$, which is called the *Lorentz factor*, quantifies how relativistic a particle is. A car moving at $100\,\text{km/h} \simeq 60\,\text{mph}$, has a Lorentz factor that differs from one by a tiny amount ($\gamma - 1 = 5.8 \times 10^{-13}$) so it is distinctly non-relativistic. Typically, any body that has $\gamma - 1 \gtrsim 0.1$ is considered to be moving relativistically.

We now list some situations in which we need to use relativistic theory.

Radio galaxies   Much of the radiation we detect from a radio galaxy is generated by electrons that have Lorentz factors $\gamma \sim 10^5$. The motion of these hyper-energetic electrons is mostly random, but in the core of a radio galaxy there is often a jet in which there is a systematic flow of plasma in which the bulk kinetic energy of the plasma is several times larger than its rest-mass energy.

Micro-quasars   These objects are essentially scaled down versions of radio galaxies: a black hole that drives a micro-quasar has a mass of a few solar masses (page 68) while a black hole in a radio galaxy might have a mass of a few hundred million solar masses. Scaling down the mass makes the phenomena physically smaller and more rapidly fluctuating, but does not change the characteristic velocities. Hence relativity is as important for a micro-quasar as for a radio galaxy.

Gamma rays   Positrons, the anti-particles of electrons, are an essentially relativistic phenomenon—P.A.M. Dirac predicted their existence while trying to make quantum mechanics consistent with relativity. When an electron annihilates with a positron, all the energy of the two particles is converted into photons, usually two photons. If the electrons are non-relativistic (so $\gamma \simeq 1$) each photon has the rest-mass energy of an electron, 511 keV (kilo-electron-volt). Gamma-ray telescopes have detected this spectral line from the direction of the centre of our Galaxy, implying the existence of a significant density of positrons there.

In 1963 the UK, the USA, and the Soviet Union signed a treaty banning tests in the atmosphere of nuclear devices. Neither side trusted the other and the USA and the USSR launched top-secret satellites that would detect gamma rays emitted by elicit tests. To everyone's surprise *many* bursts of gamma rays were detected.

The bursts lasted from seconds to a minute, and they occurred too often to be plausibly generated by nuclear devices.

After the military experts on both sides had puzzled over the data in secret, each side learnt that the other saw these events, and it became clear that the sources were astronomical. In 1973 the data were made public and it was the turn of the astronomers to be puzzled. The events seemed to be uniformly distributed over the sky, which indicated that their sources were not associated with stars in our Galaxy as most X-ray sources are. The sources had to be either within ~ 0.1 kpc of the Sun or spread through a volume much bigger than our Galaxy. But the timescales of the sources were much too short for them to be associated with active galaxies, and nobody could come up with a credible source close to the Sun. In 1986 Bohdan Paczynski had the courage to posit that, despite their small timescales, they *are* at cosmological distances, and probably associated with some kind of exploding star. In 1997 this conjecture was proved correct when the William Herschel telescope took photographs of the region around a burst that had just been detected, and the rapidly fading optical *after-glow* of the event was seen in a distant galaxy. Since then optical after-glows have been routinely detected, and we have optical spectra of the underlying objects. These data establish that many gamma-ray bursts are indeed associated with exploding stars. It has also emerged that there is more than one kind of source of gamma-ray bursts, and our understanding of these objects is incomplete. What is certain is that relativity plays an essential role in understanding these extraordinary objects.

Cosmic rays   The Earth is constantly bombarded by relativistic particles—the detection of such particles is the oldest branch of high-energy astronomy. Fortunately for us, most of these dangerous particles collide with an oxygen or nitrogen nucleus high in the atmosphere. That nucleus is badly damaged by the impact and shards flash downwards, smashing into other nuclei as they go. So a single energetic particle entering the atmosphere creates a cosmic-ray *shower*.

The particles hitting the Earth have a variety of energies but there are many more low-energy particles than high-energy ones. The most energetic particles are seen only rarely even by the detectors that have the largest collecting areas. Nonetheless, the most energetic particles so far detected have $E \sim 10^{20}$ eV, so if they are electrons they have $\gamma \sim 10^{14}$ and if they are protons or neutrons they have $\gamma \sim 10^{11}$. These energies are way higher than those achieved by the most powerful particle accelerators: the Large Hadron Collider in Geneva currently accelerates protons to $\gamma \simeq 10^3$.

Neutron stars The escape speed from the surface of a neutron star is only $\sim c/2$, so these objects are only mildly relativistic. But they lend themselves to precision measurements, so the modest relativistic effects can be precisely quantified and provide strong tests of relativity theory. Of particular importance is the *Hulse-Taylor* binary pulsar PSR B1913+16, a pair of neutron stars in orbit around one another with eccentricity $e = 0.62$ and period of 7.75 hours. This binary was discovered by Joe Taylor and his graduate student Russell Hulse in 1974, and has been intensively observed since that date. A few similar binary neutron stars have been discovered since, but the constraints they place on relativity theory are weaker because they have not been observed for so long.

X-ray sources We saw in Chapter 4, 'Accretion discs', that the innermost radii of the accretion discs around black holes are so hot that they radiate most strongly in X-rays. These regions are too tiny to be resolved by X-ray telescopes, but relativistic effects modify the shapes of X-ray emission lines that we detect.

The solar system The earth orbits the Sun at a speed of ~ 30 km s^{-1}. Since this speed is $\sim 10^{-4}c$ and relativistic effects tend to be smaller than Newtonian ones by a factor $\sim (v/c)^2$, you would expect relativity to have a very small impact on the solar system. Nonetheless, since the solar system lends itself to precision measurement, the data require relativity for their interpretation and provide crucial tests of the theory of relativity. Moreover, as we saw in Chapter 5, 'Extra-solar systems', the dynamics of the solar system is delicate and if relativistic effects were absent, the configuration of the solar system would be qualitatively different from what we actually see.

The Universe   Conceptual problems make it impossible to develop a persuasive model of the dynamics of the Universe within Newtonian physics. Hence cosmology was opened up as a branch of physics by Einstein's theory of general relativity. In the 1960s cosmology was put on a solid empirical basis by the discovery of the *microwave background radiation*, which allows us to study the Universe as it was only 100,000 years after the Big Bang, and the discovery of quasars, most of which are in galaxies that are receding from us at relativistic speeds.

Special relativity

On page 4 I explained that physicists are determined that the laws of nature shall be the same everywhere and at all times—any change in the measured phenomena from place to place or from time to time *must* be traced back to some change in the conditions under which the universal laws should be applied. The special theory of relativity articulates a new requirement for invariance: the laws shall be same in all galaxies, no matter how fast they move with respect to one another, and in all spaceships, no matter how speedy.

In 1899 Henrik Lorentz discovered a symmetry in Maxwell's equations of electrodynamics. We now call this symmetry *Lorentz covariance* and consider it to be a fundamental principle of physics, but Lorentz was unclear what the physical significance of this new symmetry was. In 1905 Einstein argued that Lorentz's symmetry reflected the fact that electromagnetism works in exactly the same way in any spaceship, regardless of the spaceship's velocity. This assertion surprised Einstein's contemporaries because Maxwell's electromagnetic waves must be propagating through some medium, the *aether*, and the aether could not be at rest with respect to all spaceships. Indeed, the velocity of spaceship Earth changes by $\sim 60\ \mathrm{km\ s^{-1}}$ every six months, and experiments with light should be able to detect such a change. Einstein's contemporaries were puzzled that no such experiment had yet succeeded in detecting our motion with respect to the aether. Einstein argued that on account of Lorentz's symmetry, it is in principle impossible to detect motion with respect to the aether. This medium, which we now call the *vacuum* has the remarkable property of looking the same to all observers, no matter how they move with respect to one another. In particular, no observer can be said to be absolutely at rest, so all motion is relative to some other observer—hence the name of Einstein's theory.

The key to getting to grips with relativity is to analyse every physical situation as a series of *events*. An event happens at some place *and some time*. So an event is specified by four numbers, the $x, y,$ and z coordinates of its place, and the time, t, at which it occurred. An observer O′ who moves with respect to the observer O who uses the four numbers (x, y, z, t) will assign a different set of numbers (x', y', z', t') to the same event. A simple rule, the *Lorentz transformation* discovered by Lorentz, enables one to calculate the primed numbers from the unprimed numbers given the velocity **v** at which O′ moves with respect to O.

The extraordinary thing about a Lorentz transformation is how it treats two events (x, y, z, t) and (x_1, y_1, z_1, t_1) that are simultaneous for O, so $t_1 = t$: in general they are *not* simultaneous for O′ in the sense that $t' \neq t_1'$. Hence O′ considers that one event happened before the other even though O knows for a fact that they occurred simultaneously. This idea that simultaneity is observer-specific is very counter-intuitive and hard to get used to.

It underlies an absolutely bewildering principle: moving clocks run slow. For example, Bob standing on a station platform studies carefully the clock on the laptop of Alice on an express as the train whizzes by. He concludes that the clock is running slow because it's a moving clock. Meanwhile Alice studies Bob's watch and concludes that his watch is running slow because it's moving with respect to her. Actually the clock is keeping perfect time when checked by Alice, and the watch likewise runs perfectly in Bob's eyes, but both run slow when checked by a moving observer. The factor by which the timepieces run slow is the Lorentz factor γ associated with the train's velocity.

Muon lifetimes

Cosmic rays detected on Earth provide a direct confirmation of the moving-clocks principle. Muons are elementary particles akin to electrons that are highly unstable so they rapidly decay into other particles—their *half-life* (the time required for half of a sample to decay) is $2.2\,\mu$s. Muons are created when cosmic ray particles hit an atomic nucleus ~ 20 km up in the Earth's atmosphere. Even moving at the speed of light, they can only travel ~ 660 m in a half-life. Hence you might think that very few, if any, of the muons created 20 km above the ground would reach the ground, contrary to what was found by sending detectors up in balloons. Relativity resolves this paradox by stating that while our clock moves through $2.2\,\mu$s, the muon's own clock advances by $2.2/\gamma\,\mu$s, so our clock must advance by $2.2\gamma\,\mu$s before the muon decays, and a muon that moves at speed $\sim c$ travels 660γ m before it decays. Hence muons created in the upper atmosphere with $\gamma \gg 1$ have a good chance of reaching the Earth.

Rest-mass energy

Alice and Bob won't agree on the energy or momentum of a given particle, because if the particle is stationary in Alice's frame, Alice will say it has no kinetic energy and no momentum, while Bob will say it has both kinetic energy and momentum. So there has to be some rule for deducing the energy and momentum that Bob assigns from the values assigned by Alice. Wonderfully, this rule turns out to be a Lorentz transformation. So if (p_x, p_y, p_z, E) are the components of momentum and the energy of a given particle of mass m_0 as seen by Alice, then we can compute the momentum and energy (p'_x, p'_y, p'_z, E') seen by Bob by applying the Lorentz transformation for the velocity of Bob with respect to Alice to the four numbers $(p_x, p_y, p_z, E/c^2)$ rather than the coordinates (x, y, z, t) of an event. This simple rule implies that the energy of a particle of rest mass m_0 is $E = \gamma m_0 c^2$. In particular, when $v \ll c$ and $\gamma \simeq 1$, we $E = m_0 c^2$, Einstein's famous expression of the equivalence of mass and energy.

According to quantum mechanics, a photon of angular frequency ω, wavelength λ that moves along the unit vector **n** has energy $\hbar\omega$, and momentum $\hbar\mathbf{k}$, where $\hbar$ is Planck's constant divided by 2π and **k** is the *wavevector* $\mathbf{k} = (\omega/c)\mathbf{n}$. When we apply a Lorentz transformation to the four numbers (k_x, k_y, k_z, ω) used by Alice, we obtain the wavevector $\mathbf{k}'$ and angular frequency ω' of the photon in the eyes of Bob. Since the photon is really a train of waves, we expect the frequency ω' measured by Bob to be *Doppler-shifted* with respect to that ω measured by Alice, and the Lorentz transformation provides the right way to compute the shift. It also enables us to discover how the direction in which the photon travels is affected by the motion of Bob with respect to Alice because these directions are given by the unit vectors $\mathbf{n} = (c/\omega)\mathbf{k}$ and $\mathbf{n}' = (c/\omega')\mathbf{k}'$.

Thus different observers disagree about the directions in which a given photon moves. An example should make it physically obvious that there will be disagreement. In an old-style western movie a guard on a moving train that is carrying gold is firing his rifle at bandits. If the guard fires perpendicular to the car, the bullet will not move perpendicular to the track because in addition to its velocity down the barrel it shares in the train's forward motion. So if the guard wants to hit a target that lies at right angles to the track at his present location, he needs to aim backwards from perpendicular to the car, so there is a component of the bullet's velocity down the barrel that cancels the forward motion of the train.

Now imagine a panicked posse of guards on the train who spray bullets uniformly in all directions. Then half of their bullets will be fired in directions forward of perpendicular to the car, and half backward of perpendicular to the car. Consequently, viewed from the ground more than half the bullets will be forward moving: the motion of the train causes the shower of bullets to be *forward beamed*.

Jets

As we saw in Chapter 4, 'Jets', several astronomical objects emit jets of plasma. The bulk velocity of the jet may yield a Lorentz factor γ of several. Within the jet photons are emitted by a variety of processes with an angular distribution that is roughly uniformly distribute in angle when viewed by an observer travelling with the material in the jet. From our perspective, these photons, like the posse's bullets, emerge with a strong bias to the forward direction—the emission is forward beamed. In fact half of the emitted photons will be travelling in directions around the jet axis that occupy just $1/(2\gamma^2)$ of a sphere. Even for modest values of γ, this is a very small fraction of the sphere. More generally Figure 26 shows the density of photons on the sphere as a function of the angle θ between the photon's direction and the jet axis. For negligible jet velocity ($\gamma \simeq 1$) the density is one for all θ. We see that even for $\gamma < 2$ forward beaming is a strong effect. In fact, the forward concentration of the *energy* emitted by the jet is even larger than Figure 26 would suggest because the photons received in the forward direction have their frequencies, and thus their energies, most strongly raised by the Doppler effect.

In general we expect an object that's prone to jet formation to emit a pair of oppositely directed jets. Generally one jet will have a

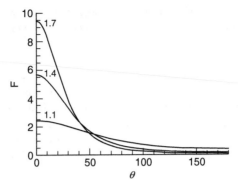

26. **The angular distribution of photons emitted by jets with Lorentz factors $\gamma = 1.1, 1.4,$ and 1.7. The quantity plotted is the the number of photons per unit solid angle at the given angle. For an isotropic distribution this is unity.**

component of velocity towards us, and the other away from us. On account of forward beaming, the approaching jet will be brighter than the receding one. Since an object has to exceed a critical brightness to be detected at all, we may detect only the approaching jet. This possibility is likely if we happen to view the object from close to the line along which the jets are being fired. Many radio galaxies appear to have only one jet.

Jets sometimes display *superluminal motion*, which is forbidden by relativity. The phenomenon is for blobs to be seen within a jet that move across the sky at a speed that seems to exceed the speed of light. The speed in question is evaluated by assuming that the blobs are moving within the plane of the sky on the ground that any component of velocity perpendicular to the plane of the sky will only increase the distance travelled, and therefore the speed derived.

Figure 27 shows the relevant geometry. A single blob is shown by an open circle twice, once at time t' (upper circle) when the blob is close to the source, and at a later time t when it is further from the

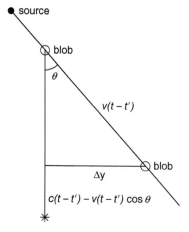

27. The geometry of superluminal expansion.

source. At both t' and t the blob emits photons towards Earth and the location at time t of the first of these photons is shown by a star at the bottom. This photon has a head start on the second photon in the race to Earth by distance $c(t - t') - v(t - t') \cos \theta$ so it will arrive at Earth sooner by

$$\Delta t = (t - t')\left(1 - \frac{v}{c} \cos \theta\right).$$

Between emitting the two photons the blob has moved a distance $\Delta y = v(t - t') \sin \theta$ over the sky. So the blob's apparent speed is

$$v_{\text{app}} = \frac{\Delta y}{\Delta t} = \frac{v \sin \theta}{1 - (v/c) \cos \theta}.$$

Figure 28 shows v_{app} as a function of θ for three values of v/c: 0.2, 0.71, and 0.96. It shows that superluminal expansion is possible for $v > 0.71c$ and that for $v \gtrsim 0.95c$, v_{app} can be several times c. Superluminal motion has been observed in many radio sources.

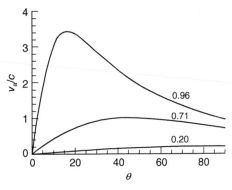

28. The apparent velocity of a blob that moves at speed $v = 0.96c$ or $0.71c$ or $0.20c$ along a line that is inclined at angle θ to the line of sight.

Shocks and particle acceleration

The space surrounding a relativistic object that fires jets is never completely empty. As the jet emerges from the source, its density may be so high that the density of the ambient medium is negligible, but as the jet moves away from the source it spreads laterally and its density declines, so eventually the ambient medium will have a significant impact on the jet.

The easiest way to imagine what happens when a jet ploughs through the ambient medium is to imagine you are moving at a velocity that is intermediate between that of the jet and the ambient medium. Let's say the jet is approaching from your left and the ambient medium is approaching from your right. The space where you stand is filled by material that was produced by the collision of jet material on ambient material. This material is incredibly hot because the ordered kinetic energy of the jet coming from the left and of the ambient material coming from the right has been randomized into frenetic darting here and there by individual particles: individual particles are zipping along, but on the average they are going nowhere, so this *shocked plasma* is at rest with respect to you.

The region of shocked plasma moves away from the source and grows steadily as a result of fresh jet material hitting it from the left and ambient material hitting it from the right. The narrow regions in which the systematic motions of the jet and ambient particles are randomized are called *shocks*.

Typically the shocks are what physicists call *collisionless* shocks. Given that a shock is precisely a place where fast-moving fluid collides with slower fluid, the collisionless sobriquet sounds daft. What it means is that the velocities of particles are changed by an electromagnetic field that has a much longer lengthscale than the inter-particle separation. Thus an incoming electron or proton is not decelerated by colliding with an individual atom or ion, but by a fairly smooth electromagnetic field generated by zillions of electrons and ions collectively (Figure 29).

The origin of this field is fundamentally separation of electrons from ions arising from the enormous mass difference (by a factor in excess of 1,800) between electrons and ions: incoming electrons decelerate much sooner than the more massive ions, so regions of

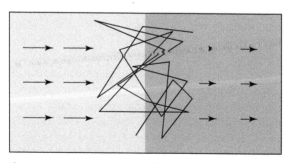

29. **Plasma coming fast from the left hits slower-moving plasma on the right. On impact the plasma slows down and becomes denser, symbolized the by the smaller arrows and darker shading. The zig-zag line shows the trajectory of a fast particle that is multiply scattered, often crossing the interface between the two regions between scatterings.**

positive and negative charge density arise. These regions generate an electric field which pulls one way on the electrons and the opposite way on the ions, thus tending to bring them to the same mean velocity. Differences in the mean velocities of electrons and ions imply the existence of electric currents, and these currents generate a magnetic field. Moreover, the flows of electrons and ions is highly unsteady in this region, so the electric and magnetic fields are time dependent. Because the fields are time-dependent, they can change the energies of individual electrons and ions, and on the average they transfer energy from the ions to the electrons – in the original, ordered flow upstream of the shock, both species had the same velocity, so the kinetic energy of the material was overwhelmingly contained in the ions. The post-shock plasma is *relaxing* to a state of thermal equilibrium in which each species has precisely half the now randomized kinetic energy. Net transfer of kinetic energy from the ions to the electrons is a key part of this relaxation process.

The shocked plasma is analogous to a gambling den into which punters bring money which is redistributed in the den. We've just discussed how this transfer affects the average punter. But a few punters grow extraordinarily rich as a consequence of first becoming unusually rich.

In the plasma the analogue of wealth is kinetic energy, and the faster a particle moves, the harder it becomes to deflect. Particles that become sufficiently fast can crash right out of the shocked region, and thus enter one of the regions to right or left in which there is an ordered inflow. Since these regions are very extensive, the particle *will* eventually be deflected there and find its way back to the shocked region. But when it returns it will be moving faster than when it departed because the net effect of its deflections in the ordered flow is to reverse the sign of its velocity *with respect to the flow*. In the non-relativistic case its speed is now the sum of its original speed and the speed of the inflowing material. Since the particle is now moving faster than ever, it is likely to crash right

100

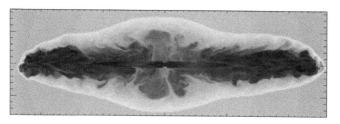

30. Computer simulation of a jet-inflated cocoon. Dark shading indicates low density. The grey at the edge represents the undisturbed circumgalactic gas. The light band inside this is the result of compressing this gas in a shock. The dark shades further in show extremely hot, turbulent plasma that was heated in the shocks at the ends of the jets.

through the shocked region and enter the opposite region of ordered inflow, where again the sign of its velocity with respect to that flow will be reversed and it will return at an even greater speed. By this process of *Fermi acceleration* particles can acquire very large Lorentz factors. In fact this is how the cosmic rays we detect on Earth are accelerated.

Because it is very hot, the shocked plasma is a high-pressure fluid and expands in any direction it can. The *ram pressure* of the jet to the left and of the ambient medium to the right prevent it expanding in these directions, but it can usually expand in the perpendicular directions. Flowing outwards perpendicular to the jet in this way it gradually inflates a *cocoon* of plasma that surrounds the jet as sketched in Figure 30.

Synchrotron radiation

Within the cocoon there are many electrons that were accelerated to large Lorentz factors by the Fermi process. The cocoon invariably contains a magnetic field, and the electrons spiral around the field's lines of force, emitting electromagnetic radiation as they do so. If the electrons are non-relativistic, the radiation is all at one frequency, the *Larmor frequency* ν_{L}, which is proportional to the strength of the magnetic field. If the electrons are relativistic, the radiation covers a band of frequencies that extends up to $\gamma^2 \nu_L$ and the radiation is called *synchrotron radiation*. The typical Lorentz factor of the emitting electrons can be inferred from the spectrum of the radiation. Because magnetic fields in interstellar and intergalactic space are generally weak, radio telescopes are only sensitive to synchrotron radiation from electrons with Lorentz factors $\gamma \gtrsim 10^4$, yet sychrotron radiation is often observed.

General relativity

We have seen that the special theory of relativity emerged from Maxwell's equations of electrodynamics, but it laid bare a fundamental symmetry of space-time. Einstein became convinced that *all* fundamental physical laws should display this symmetry, and a theory that conspicuously failed to do so was Newton's theory of gravity.

There is an extremely close analogy between gravity and electrostatics: like gravity, the electrostatic force is proportional to the inverse square of distance. The theory of relativity revealed that magnetism is a relativistic correction to electrostatics in the sense that a moving observer sees an electric field in part as a magnetic field, so we expect a gravitational field to look rather different to a moving observer. In particular, we should expect the complete gravitational force on a body to depend on its velocity, just as the electromagnetic force on a charge has a magnetic component that's proportional to velocity. Since photons move faster than any particle of non-zero rest mass, understanding this component must be essential if you want to understand how a gravitational field affects photons.

There's a special aspect to gravity that electrostatics lacks and Einstein was convinced was of fundamental importance. This is that the gravitational force is proportional to mass. Legend has it that Galileo demonstrated this fact around 1600 by dropping two balls with very different weights from the top of the Leaning Tower of Pisa: despite their different weights, the balls landed essentially simultaneously. Galileo's experiment was not very precise. A much better test of the dependence of the gravitational force on mass is to measure the periods of pendulums that have the same length but bobs made of different materials and masses. In 1891 Baron Roland Eötvös devised an experiment that demonstrated the proportionality to extremely high precision by probing the balance between an object's gravitational attraction to the Sun with the force required to keep it in orbit around the Sun. Einstein considered that such a precise proportionality couldn't happen by chance; it must emerge as inevitable from the correct theory of gravity.

Einstein struggled for ten years to put mathematical flesh on this idea, and the theory he produced is undoubtedly one of the greatest creative achievements of mankind. In essence, he did for gravity what Maxwell had done for electromagnetism, namely draw disparate bits of physics into a single coherent mathematical structure, which contained not only the physics that inspired it but also predictions of entirely new phenomena.

An electromagnetic field is generated by the density of electric charge and current. A gravitational field is generated by the density of energy-momentum and the fluxes of this quantity. In Newtonian physics gravity is generated by mass, which through $E = mc^2$ is equivalent to energy. General relativity teaches that this view is too limited, and arises because we haven't had experience of fast-moving massive bodies or very high pressures, so we are unaware that both momentum and a flux of energy-momentum also help to generate the gravitational field. We also think of gravity as an intrinsically attractive force, while general relativity shows that it can equally be repulsive.

Maxwell's equations are differential equations that can be solved for the electromagnetic field generated by a given density of electric charge and current, and Einstein's equations are differential equations that can be solved for the gravitational field that is generated by a given density and flux of energy-momentum. Whereas Maxwell's equations are linear, Einstein's equations are *non-linear equations*: if an equation is *linear*, you can find simple solutions and add them together to build more complex solutions. When an equation is non-linear, the sum of two solutions is generally not a solution, so you cannot build up sophisticated solutions by adding simple ones.

On account of this problem the only exact solutions of Einstein's equations that we have are ones with special symmetries. The first and most famous of these solutions is that found in 1916 by Karl Schwarzschild. This solution describes the gravitational field that surrounds a spherical mass, and in Chapter 5, 'Extra-solar systems', we discussed its application to the solar system. In 1963 Roy Kerr extended this solution to the gravitational field that surrounds a spinning body, and this solution is important for understanding the inner regions of accretion discs (Chapter 4, 'Journey's end'). Other exact solutions describe homogeneous universes. A exact solution from this rather limited set can be extended to an approximate solution by perturbation theory (cf. Chapter 5, 'Dynamics of planetary systems'), and many astronomical applications of general relativity rely on this approach. Since 2000 critical advances have been achieved in the numerical solution of Einstein's equations. These advances arise in part from the steady growth in available computing power, but they are mainly due to better understanding of how the equations need to be tackled.

Weak-field gravity   Two conditions have to be satisfied for Newton's theory of gravity to be accurate: first, the gravitational field must be weak and speeds must be much less than c. The second condition is the more restrictive in practice, not least because photons always violate it. In astronomy the field is usually weak and time-independent. Then the gravitational field is completely described by the Newtonian gravitational potential Φ—for example, distance r from a point mass M we have, $\Phi(r) = -GM/r$.

Gravitational redshift

Suppose we measure the frequency of a spectral line emitted by atoms on the surface of a compact object such as a white dwarf. Then standing at $\mathbf{x}_\mathrm{o}$ we are effectively listening to the ticks of an atomic clock that does not move relative to us and sits at $\mathbf{x}_\mathrm{c}$ near the compact object. Suppose the clock ticks once a second. Then the time that elapses between our receiving light pulses sent out with each tick is

$$t_\mathrm{o} = \sqrt{\frac{1 + \frac{1}{2}\Phi(\mathbf{x}_\mathrm{o})/c^2}{1 + \frac{1}{2}\Phi(\mathbf{x}_\mathrm{c})/c^2}}\ \mathrm{s}.$$

Since the clock is closer to the compact object than we are, $\Phi(\mathbf{x}_\mathrm{c}) < \Phi(\mathbf{x}_\mathrm{o})$ and we have $t_\mathrm{o} > 1\,\mathrm{s}$. That is, we think the clock is running slow. Reverting to the case in which the clock is an oscillating atom, we have that the measured frequency $\nu_\mathrm{o} = 1/t_\mathrm{o}$ is lower than the intrinsic frequency $\nu_\mathrm{c} = 1/t_\mathrm{c}$, where t_c is the time between ticks as measured at $\mathbf{x}_\mathrm{c}$. This is the phenomenon of *gravitational redshift*.

Gravitational lensing

Light travels through glass or water slower than it does through air, with the consequence that rays of light are bent when they enter glass or water. This phenomenon is usually quantified by the *refractive index n*, which is defined such that the speed of light in a medium of refractive index n is c/n.

If we use ordinary Cartesian coordinates (x, y, z) and assume that the distance between two points x_1 and x_2 on the x axis is $s = x_2 - x_1$, then a gravitational field endows the vacuum with a refractive index

$$n = 1 + \frac{|\Phi|}{2c^2}. \tag{6.1}$$

With the rather natural definition of distance that we've adopted, when $\Phi \neq 0$ light appears to travel through the vacuum slower than the speed of light. Actually it always travels exactly *at* the speed of light and our expression gives a lower value because we have under-estimated the distance between x_1 and x_2. But it is very useful to imagine that photons travel slower than c when $\Phi(\mathbf{x}) < 0$.

This conceit is helpful because it allows us to use our knowledge of optics to predict how light will be deflected by a gravitational field. A lens brings a parallel beam of light to a focus by slowing photons that pass through the centre of the lens more than it slows photons that pass far from its axis, where it is thinner (Figure 31, upper panel). In fact, a lens is designed such that the time taken by photons to travel from a distant source to the image is

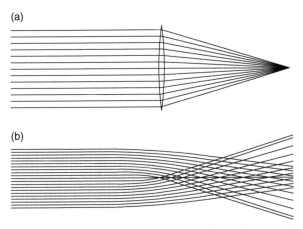

31. A lens is designed to cause a bundle of initially parallel rays to all pass through the focus (a). A gravitational lens (b) causes initially parallel rays to converge but does not cause them to pass through a single point. The mass of the lensing object is unrealistically massive by many powers of ten so the deflections can be easily seen.

independent of the distance from the axis at which a photon passes the lens—this is *Fermat's principle of least time*.

Sometimes our line of sight to a distant object such as a quasar passes close to the centre of a massive object such as a galaxy or a cluster of galaxies. Then the gravitational field of the intervening object acts like a lens in causing rays that were diverging from the source to converge on a point near us. The lower panel of Figure 31 shows a quantitative example of this phenomenon.

As Figure 31 illustrates, the lens formed by a typical gravitational field is of poor quality in that it does not cause all rays to cross at a point, and consequently it produces out-of-focus and distorted images. In fact, a gravitational lens is liable to form several images of the same object. In 1979 the first example of this *strong lensing* phenomenon, SBS 0957+561, was discovered, and it remains one of the most dramatic examples: a galaxy at redshift $z = 0.355$ forms two images of a single quasar 6 arcseconds apart. Since the discovery of SBS 0957+561 systematic searches for multiply imaged quasars have been conducted and hundreds of examples have been found, some with four images, but almost always smaller image separations than that of SBS 0957+561.

As we saw in Chapter 4, 'Time-domain astronomy', the luminosities of quasars fluctuate quite significantly on timescales of months or years. When a quasar is multiply imaged, each image is associated with a particular time of passage of photons from the quasar to Earth. The differences in these times can be measured by finding the time by which one has to shift forwards or back a record of the brightness of one image to make it match the brightness record of another image. These time delays can be computed for any model of the gravitational field that forms the lens, and they are proportional to the *Hubble constant* (*H*), which relates redshift (*z*) to distance (*s*) through $z = Hs/c$, because this determines the distance to an object of known cosmological redshift: the bigger this distance is, the longer is the journey the

32. **Weak lensing in the galaxy cluster Abell 2218. The images of galaxies that lie behind the cluster are stretched by the cluster's gravitational field into arcs perpendicular to the field direction.**

photons have taken to reach us, and the greater is the time delay associated with choosing a longer route.

Multiple imaging of quasars and galaxies is a rare phenomenon. What's extremely common is for the gravitational field along lines of sight to distant galaxies to distort and magnify their images (Figure 32). This phenomenon is called *weak lensing*. In weak lensing the gravitational field plays the role of badly polished primary lens of a gigantic telescope. Weakly lensed galaxies may be sufficiently brightened by the lens for it to be possible to study them in much greater detail than normal galaxies at their redshift.

Astronomers can even take advantage of the poor quality of gravitational lenses. A weakly lensed image is stretched perpendicular to the direction of the gravitational field. So if we viewed a population of intrinsically round galaxies through a gravitational lens, the galaxies would appear elliptical. The short axes of the ellipses would be parallel to the projection onto the sky of the gravitational field, and the axis ratios of the ellipses would indicate the strength of the field. Using this idea to measure gravitational fields is hard in practice, not least because galaxy images are intrinsically elliptical. However, distortion of images by the gravitational field tends to align the ellipses of galaxies that are close on the sky, and precision measurement of this effect is now a major tool for cosmology.

Gravitational micro-lensing

When a foreground star passes in front of a background star, the gravitational field of the foreground star may strongly lens the background star. Usually the images of the background star form so close together that current telescopes cannot separate them, but lensing can nonetheless be inferred from the way the background star brightens as the gravitational field focuses its light on the Earth. The formation of unresolvable multiple images is called *micro-lensing* and constitutes an important probe of our Galaxy's content. Often the foreground star is too faint to be detected, so all we see is a single star temporarily surging in brightness.

Since the late 1990s the brightnesses of hundreds of millions of stars have been monitored every night and thousands of instances of micro-lensing have been discovered. Interpretation of the data is difficult because many stars have fluctuating luminosities. There are two ways to distinguish luminosity variations from micro-lensing: (i) the former is generally associated with a colour change while micro-lensing is not; and (ii) a single star has a negligible probability to be micro-lensed more than once in recorded history, while a star that fluctuates in luminosity once is very likely to do it often.

Micro-lensing is important because it allows us to detect the gravitational fields of objects that are far too faint to be directly seen. In fact, the probability that any given star is being micro-lensed at any given time depends only on the mass density of the lensing objects along the line of sight to the star, and not on the mass of an individual object. So the probability that a given star is being micro-lensed tonight might be 10^{-6}, regardless of whether the mass density is made up of black holes of mass $1,000 \, M_\odot$, or stars of mass $1 \, M_\odot$, or Jupiters of mass $10^{-3} \, M_\odot$. What changes between these cases is the duration of a typical micro-lensing event, which is proportional to the mass of the lensing object. So if black holes are responsible for lensing, each event will last a million times longer than if Jupiters do the job, and there will be a million times fewer events per year. Hence by monitoring the brightnesses of stars, one can determine both the density of the lensing objects and their typical mass. If the lensing objects are too massive, we'll have to be very lucky to detect a single event, and if their masses are too small the events will be so brief that we won't have enough observations during any given event to detect the characteristic rise and fall in brightness that distinguishes a microlensing event from noise. But the range of masses to which observations are sensitive is very large $\sim 100 - 10^{-3} \, M_\odot$. Thus microlensing has been used to place upper limits on the space density of very low-mass stars and free-floating planets within our Galaxy. It has also been used to detect planetary systems that could not otherwise be detected.

When a routinely monitored star begins to brighten in a way suggestive of micro-lensing, a model of the lens is fitted to the data. If this model suggests that the background star is going to pass very close to the centre of the lens (where an unseen star sits), observers all round the world, including many amateur astronomers, are alerted because it is then key to keep the star under twenty-four-hour surveillance, and this cannot be achieved from just one or two sites. In the few hours that the star is very close to the centre of the lens, the contributions of planets to the lensing gravitational field can modify the measured brightness quite significantly, and thus betray their existence (Figure 33).

Deflection of light by the Sun

The gravitational field of the Sun forms a lens in which we are embedded. At our distance from the Sun, equation (6.1) gives a

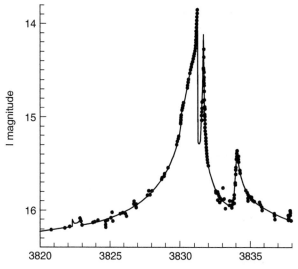

33. The micro-lensing event OGLE-2006-BLG-109. The brightness of the star measured at 12 observatories is plotted against time in units of a day. The gravitational fields of two planets generate extremely rapid brightness fluctuations. These data yield the mass of the star as $0.51\,\mathcal{M}_\odot$ and the planets' masses as $231\,\mathcal{M}_\oplus$ and $86\,\mathcal{M}_\oplus$ similar to Jupiter and Saturn.

refractive index that is very close to unity ($n - 1 \simeq 5 \times 10^{-9}$) so light rays are bent through only tiny angles on their way to us unless they pass quite close to the surface of the Sun, where $n - 1$ is $\sim$ 100 times larger. Deflection of the light from stars by the Sun's gravitational field shifts the position of each star on the sky, and these shifts are in principle detectable by comparing images of a given star field taken at different times of the year.

When Einstein's theory was still new and untested, stars could be observed close to the Sun only during a solar eclipse. Moreover, the required measurement was extremely challenging as even for a star that is seen at the edge of the Sun's disc, the predicted shift in position is only 1.75 arcsec and all the stars in a small field of view will be shifted to a similar extent—measuring the absolute positions of stars is very much harder than measuring the angles between neighbouring stars. Nonetheless, during the 1919 eclipse a team led by Arthur Eddington measured shifts that were consistent with Einstein's prediction.

From a spacecraft stars can be observed even close to the Sun, but space telescopes avoid doing this because their delicate detectors would be fried if by chance any part of the Sun's disc came into the field of view. The Gaia satellite, launched in December 2013, can measure the positions of stars to such exquisite precision ($\lesssim 0.00001$ arcsec) that allowance must be made for Einstein's shift over the whole sky. In fact, the analysis also has to take acount of the deflection of light by planets.

Shapiro delays

In the 1960s it became possible to bounce radar waves off planets. The idea was to measure the time it took a pulse to return to Earth and to compare it with the time predicted by a general-relativistic model of the solar system. Two difficulties with these early experiments are that (i) reflection off a planet does not occur instantaneously, but over a period that depends on the planet's shape, and (ii) radio waves do not travel through interplanetary space at precisely the speed of light because space contains a tenuous plasma, which shifts the refractive index from unity. Both of these problems could be eliminated by replacing the planets with spacecraft programmed to respond to an outgoing signal by returning a signal after a precisely known delay—the impact of intervening plasma can be determined by comparing results obtained using different transmission frequencies because the delay caused by the plasma is frequency-dependent.

These experiments directly probe the gravitational field within the solar system, which is expected to be a slightly perturbed form of Karl Schwarzschild's solution of Einstein's equations. The conclusion is that any difference between the true and predicted fields must be smaller than parts in a thousand.

Pulsars and gravitational waves

Many, perhaps all, neutron stars have magnetic fields sticking out of them, that sweep the surrounding space as the neutron star spins. The rotating magnetic field can cause a beam of radio waves to sweep through space very much as the rotating lantern of a lighthouse sweeps over the ocean. The periodic passages of a beam over the Earth generates the highly characteristic radio signal of a *pulsar*.

A neutron star spins in a very regular way because it's hard for anything in its environment to apply a significant torque to it. So precision measurements can be made by comparing the times at which pulses are received with the times at which we calculate they were emitted, given the steady rotation of the neutron star. The most interesting object from this point of view is the Hulse–Taylor pulsar (page 90). The distance between the stars varies between 0.75 and 3.15 million kilometres (for comparison the radius of the Sun is 0.70 million kilometres). One of the neutron stars is a pulsar and general relativity predicts quite a complex pattern of arrival times for its radio pulses because the distance each pulse has to travel to reach us is constantly changing, as is the effective refractive index (see equation (6.2)) of the space the pulse has to traverse as it moves through the intense gravitational field of the binary neutron star. The predicted pattern is in excellent agreement with the measurements.

Any theory of gravity that is consistent with the symmetry that Lorentz's transformation uncovered will predict that a binary star emits gravitational waves. Indeed, when the sources of a gravitational field move, the field will be updated to the new source positions much sooner near the sources than at distant locations, and the updating will be accomplished by waves in the field that spread out from the moving sources. The basic physics of radiation is the same for gravitational waves as it is for electromagnetic waves, so the key to effective radiation is for the source (antenna) to be not much smaller than the wavelength of the waves (cf. Chapter 2, 'Emission by gas'). In the case of a binary star this condition translates into the stars moving not much slower than the speed of light. So two neutron stars that are almost touching would radiate gravitational waves with high efficiency and lose energy on a timescale of a few orbital periods (a fraction of a second), while a binary star with a separation of $\sim 1\,\mathrm{AU}$ and a period of a year is a dreadfully inefficient radiator of gravitational waves. One of the best radiators we know is the Hulse–Taylor pulsar. It isn't a terribly good radiator: its timescale for energy loss is $\sim 0.3\,\mathrm{Myr}$, or ~ 340 million orbital periods. Nonetheless, the precision of the measurements of pulse arrivals is such that the change in the period caused by gravitational radiation has been measured and found to be in excellent agreement with the prediction of Einstein's theory.

At the time of writing gravitational waves have yet to be detected because it's hard to build an efficient antenna if the fundamental requirement is massive bodies moving near the speed of light. Detectors are being perfected in which light is passed to and fro down two evacuated tunnels 5 km in length that are at right angles to one another. Interference fringes are observed between light that has been up and down one tunnel and light that has travelled the other tunnel. When a gravitational wave passes over the system it changes the effective refractive index (see equation (6.2)) within the tunnels and thus shifts the interference fringes. The expected effect from any astronomical source is absolutely tiny, but within a few years it should be observed. This feat will arguably be the all-time toughest of experimental physics.

Chapter 7
Galaxies

When you look up at the night sky on a dark night, the points of light above you are overwhelmingly stars within our Galaxy—two or three of the brighter points will be planets, and in a very dark site you may be able to make out the faint smudges of the Andromeda nebula or, if you can see far enough south, the Magellanic Clouds. By contrast, most of the sources detected by state-of-the-art telescopes are galaxies. The Universe seems populated by galaxies in the same way that our Galaxy is populated by stars.

Galaxy morphology

To an excellent approximation a galaxy consists of a huge number of point masses that move freely in the gravitational field that they jointly generate. Some of these masses are stars, but most are thought to be elementary particles of a still unknown type, which together comprise *dark matter*: material we cannot see but detect through its gravity. Astrophysics is made much simpler by the fact that, despite their hugely discrepant masses, stars and dark-matter particles have the same equations of motion—they move non-relativistically in a common gravitational field. Their typical orbits are nonetheless different: dark-matter particles tend to be on more energetic orbits that take them further from the centre of the galaxy, and we think their orbits are less concentrated around the galaxy's equatorial plane than are those of the stars, many of which are confined to a thin disc containing that plane.

The fraction of a galaxy's mass that is contained in stars rather than dark matter varies considerably. In the least massive galaxies, *dwarf spheroidal galaxies*, less than 1 per cent of mass is contained in stars. Our own Galaxy belongs to the class of galaxies that have the largest mass fractions in stars, and this fraction is $\sim$ 5 per cent. So whatever type of galaxy you choose to consider, dark matter dominates the overall mass budget. However, in a galaxy such as ours stars dominate the mass budget within a few kiloparsecs of the centre, while dark matter dominates further out. Dwarf spheroidal galaxies, by contrast, are dominated by dark matter at all radii.

A fundamental property of the population of galaxies is the *galaxy luminosity function* plotted in Figure 34. This shows the number of galaxies per unit interval of the logarithm of luminosity $\log L$. We see that there are myriads of faint galaxies, and rather few luminous galaxies: at the left-hand, low-luminosity, side of Figure 34, the luminosity function falls as a straight line, and then near a characteristic luminosity L^* it turns strongly downward. The *Schechter luminosity* L^* coincides almost exactly with the luminosity of our Galaxy.

From the galaxy luminosity function you can ask the question, if I repeatedly pick a star at random from all the stars in the Universe, what will be the average of the luminosities of the galaxy of the chosen stars? The answer turns out to lie close to L^*, so it is probably no accident that this is the luminosity of our Galaxy: we expect there to be innumerable civilizations in the Universe that have addressed this question, and most of them are warmed by a star that lies in a galaxy like ours.

Decomposition into components

It's helpful to imagine that a galaxy comprises a few *components*. A conspicuous component of our Galaxy is the disc, within which

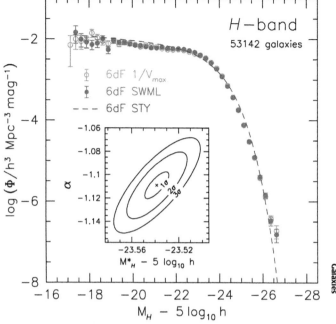

34. Galaxy luminosity function. The space density of galaxies is plotted against luminosity. Both scales are logarithmic.

the Sun lies (Figure 35). Stellar discs are found to have surface densities that fall off with radius roughly exponentially.

The stellar discs of galaxies frequently have embedded within them a gas disc like the Galaxy's gas disc, which we described in Chapter 2, 'The gas disc'. A gas disc is generally more radially extended than the stellar disc and distinctly thinner. Galaxies that possess a significant gas disc generally have spiral arms within both their gas and stellar discs. Spiral arms are generally absent when the gas disc is insignificant. Galaxies that have a prominent stellar disc but only an insignificant gas disc are called *lenticular* or *S0* galaxies. Our Galaxy is a *spiral* galaxy.

35. An image of our Galaxy constructed by counting half a billion stars. Clouds of obscuring dust are evident.

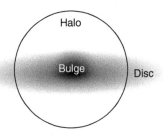

36. Cartoon showing disc, bulge/spheroid, halo.

The inner ~ 3 kpc of our Galaxy is dominated by the *bulge* or *spheroid* (Figure 36). As these names suggest, this stellar component is much less flattened towards the equatorial plane than the disc. The bulge of our Galaxy is not axisymmetric (which the disc nearly is), but forms a bar. The latter is about three times long as it is wide and its long axis lies in the Galactic plane. The bar rotates at the centre of our Galaxy like the beater of a food mixer. Meanwhile its stars circulate rather faster within the bar, moving on quite eccentric orbits. The bar excites spiral waves in the disc around it, but these waves move around the disc more slowly than the bar rotates.

In galaxies like ours the bulge is often barred. But not all bulges are barred, and not all spiral galaxies have bulges. For example, the third most luminous member of the *Local Group* of galaxies (of which our Galaxy is the second mast luminous member) is the *Triangulum Nebula* or *M33*, and it does not have a bulge. The existence of bulgeless galaxies such as M33 is currently rather puzzling for cosmologists.

Disc galaxies are ones in which the bulge is subordinate to the stellar disc (Figure 35). In an *elliptical galaxy*, by contrast, the bulge dominates the disc to the extent that the disc can be detected, if at all, only by the most minute quantitative analysis. Elliptical galaxies are usually, but not always, axisymmetric. In an elliptical galaxy the motions of stars are much less ordered than in a disc—where rotation around the approximate symmetry axis is very dominant—and rather less ordered than in the bulge of a galaxy such as ours.

Stellar dynamics

Since most of the mass of a galaxy is contained in particles, stars, and dark matter, which rarely collide, we need to understand how a large number of point masses move under their mutual gravitational attraction—a branch of astrophysics known as *stellar dynamics*. The particles in question can be stars or dark-matter particles, it makes little difference.

Is it safe to treat stars as point masses? The answer to this question is normally a resounding 'yes': about 3 Gyr from now our Galaxy will collide and merge with our nearest massive neighbour, the Andromeda Nebula. Then for hundreds of millions of years streams of 10^{11} stars from each galaxy will rush through each other as if in a stupendous military tattoo, and the number of physical impacts expected is less than one! In fact, the only environment in which treating stars as point particles may be problematic is in the immediate vicinity of a galaxy's central black hole. But even here physical contact can be significant only for giant stars, and such contact is likely merely to strip away the bloated atmosphere from such a star, without depriving it of most of its mass.

The Sun is on a near-circular orbit around the centre of our Galaxy some 8.3 kpc away. The gravitational pull towards the Galactic centre that holds it on this orbit is the sum of the pulls of the 10^{11} stars and zillions of dark-matter particles that make up the Galaxy. The fraction of this force contributed by the near neighbours of the Sun is negligible. This is a dramatically different situation from that in a solid or liquid: the force on an atom is completely dominated by the handful of nearest neighbours because the inter-atomic force declines with distance much more rapidly than the gravitational attraction between stars. Since the force on the Sun is dominated by large numbers of distant objects, the force would change very little if the Sun were moved a parsec or so in any direction, and it won't change very much in the next million years or so: the force is a very smooth function of position and time. Consequently, we can compute the orbit of a star such as the Sun to high precision by spreading the mass of each particle smoothly over a few inter-particle distances, and computing the gravitational field of the resulting continuous mass distribution. Our first step when considering any stellar system is to do just this and then to investigate the nature of orbits in the smooth gravitational field. In so far as this approach is adequate, we say the stellar system is *collisionless*.

You can specify an orbit by giving a position $\mathbf{x}$ and a velocity $\mathbf{v}$ that a star on this orbit has at some particular time. If every such pair $(\mathbf{x}, \mathbf{v})$ specified a different orbit, the space of orbits would be six dimensional, because a position and a velocity both live in three-dimensional spaces. But it is clear that different pairs $(\mathbf{x}, \mathbf{v})$ do not necessarily generate different orbits, because any pairs that occur at different times along an orbit obviously generate the same orbit.

Integrating orbits in typical galactic gravitational field reveals that the space of orbits is three-dimensional. That is, an orbit can be uniquely specified by three numbers. These numbers are called *constants of motion* because their values do not change as one moves along an orbit. The fundamental task of stellar dynamics is to learn how to compute suitable constants of motion J_i from a pair $(\mathbf{x}, \mathbf{v})$. That is, one needs an algorithm to compute three functions $J_i(\mathbf{x}, \mathbf{v})$.

The dynamical state of a galaxy can then be reduced to a density of stars and dark-matter particles in the imaginary space in which the Cartesian coordinates of points are the three numbers J_i. This is known as *action space*. Our knowledge of the action-space density of stars and dark-matter particles is still incomplete even for our own Galaxy, and is very incomplete for all external galaxies.

Galaxies, gases, and crystals

A galaxy, like a litre of gas or a diamond, consists of a large number of particles interacting with one another. Statistical physics provides a rather complete understanding of bottles of gas and crystals by starting from the concept of thermal equilibrium: the state into which the system relaxes if you leave it alone for long enough. The *principle of maximum entropy* tells us how to compute the arrangement of the system's particles when it's in thermal equilibrium. For example, in the case of a gas the principle of maximum entropy enables us to work out how many molecules are moving at any given velocity (the *Maxwellian distribution*), and what pressure the gas exerts given its energy and volume. Then by perturbing the state of thermal equilibrium, we determine the system's *transport coefficients*, such as its sound speed, thermal conductivity, viscosity, etc.

Unfortunately, the very first step in this chain of analysis is inapplicable to a galaxy, because a galaxy has no state of maximum entropy, and is therefore incapable of thermal equilibrium.

Entropy is disorder: the principle of maximum entropy simply says that in thermal equilibrium the system is as disordered as it can be given its energy, volume, and any other restrictions on rearrangements of its particles. A self-gravitating system such as a galaxy or a star can always increase its entropy by moving mass inwards to increase the intensity of the gravitational field near the centre, and then transporting outwards the energy that is released by this local contraction, and giving it to peripheral particles: this energy increases the distance from the centre to which these particles can cruise so their disorder increases. In Chapter 3, 'Life after the main sequence' we saw that late in the life of a star, its core contracts and its envelope swells up. This is because the star is increasing its entropy by the process we have just described.

Equilibrium dynamical models

The major problem we face, because galaxies can't attain thermal equilibrium, is how to deduce the basic distribution of stars and dark matter particles. Once we know what that is, we can compute the transport coefficients. But we we need to know what configuration to perturb, and we have no principle from which to deduce it. One workaround is to rely on a cosmological simulation of galaxy formation, and another is to fit a dynamical model to observational data.

Cosmological simulations do not provide useful predictions on their own because we lack the resources required to simulate the extremely complex physics of star and galaxy formation. Consequently, all simulations rely on mathematical formulae that it is hoped approximate the outcome of physics that has been left out, and the parameters in these formulae have to be calibrated against observations. So if you want to make galaxy models, you should go straight to the observations rather that bothering with simulations.

Elliptical galaxies   Elliptical galaxies are the easiest to model and significant numbers of such galaxies have been modelled dynamically. One important conclusion from these models is that most of these objects are nearly axisymmetric and flattened by their spins. However, the most massive elliptical galaxies spin very slowly and have triaxial shapes, rather like the kernel of plum. The low spins and triaxial shapes of these objects probably arise because they are the products of mergers of two gas-poor galaxies of comparable mass.

Another important conclusion from models of elliptical and lenticular galaxies is that the more luminous a galaxy is, the richer it is in heavy elements and the bigger is the proportion of its mass that is contributed by dark matter. The heavy-element richness of massive galaxies probably arises because supernovae have greater difficulty driving the products of nucleosynthesis out of the deeper gravitational potential wells of higher mass galaxies. The increasing contribution of dark matter may arise because more massive galaxies typically have lower star densities than less massive galaxies.

Models of the very centres of elliptical and lenticular galaxies are especially interesting because they may enable us to detect a central *super-massive black hole*. The key idea is that interior to the radius r_{infl} at which the black hole contributes as much to the gravitational field as do the stars, the random velocities of stars must rise roughly like $1/\sqrt{r}$. A secure detection of a black hole requires measurements of both the star density and the random velocities of stars very close to the centre. The Hubble Space Telescopy has been invaluable in gathering these data.

The inferred mass of the black hole proves to be tightly correlated with the magnitude of the random velocities of stars at radii that are significantly bigger than r_{infl}. This finding suggests a causal connection between the growth of the black hole and the growth of the galaxy's stellar population, and it has been argued that this is surprising because the stellar population is enormously more massive and physically extended than the black hole. However, the space density of quasars—black holes that are rapidly accreting cold gas—peaks at precisely the redshift $z \sim 2$ at which the cosmic star-formation rate was highest. Since stars form from cold gas, it seems likely that the growth rate of a black hole and its host stellar population both track the availability of cold gas, so it is natural that their current masses are tightly correlated.

Spiral galaxies The spiral galaxy that has been most extensively modelled is our own. Idealised equilibrium dynamical models of the thin and thick discs have been constructed, and from their vertical structure you can infer that 56 per cent of the gravitational force that holds the Sun in its orbit is generated by dark matter so only 44 per cent is generated by stars. The mass of the disc is consistent with the disc consisting exclusively of stars and gas rather than dark matter.

Slow drift

No star stays on the same orbit for the entire life of our Galaxy because the smooth, time-independent gravitational field that we assume when computing the constants of motion J_i is an idealization. The real gravitational field fluctuates around the idealized field for several reasons. First the disc supports spiral structure that is not reflected in our idealization. Second, the disc contains massive clouds of molecular gas that form, move through the disc and disperse in a random way. The gravitational fields of these objects we likewise ignored in our idealization. Third, no galaxy is isolated; other galaxies, many of them small, are constantly falling into a massive galaxy such as ours, and these objects can orbit through the galaxy for a long time before they disperse. In our idealization, we ignored these massive moving lumps. Finally, there are 'Poisson fluctuations' in the number of point masses in any volume: if the density of particles is such that on average we expect N objects in a volume V, then the number actually in that volume will fluctuate over time by $\sim \sqrt{N}$. Since the force arising from volume V is proportional to the number of masses it contains, the gravitational force will fluctuate by a fraction $\sqrt{N}/N = 1/\sqrt{N}$ of itself.

If we imagine a star or dark-matter particle to be on an orbit through the idealized smoothed gravitational field, we must consider it to be jiggled by a small random field. On account of this random field a star or dark-matter particle that's initially on the orbit $\mathbf{J}$ has a probability in a given time t of transferring to a different orbit $\mathbf{J}'$. This is closely analogous to the Brownian motion of pollen grains on the surface of water: these are seen to jiggle around at random, so in some short time t they have a probability to move from position $\mathbf{x}$ to some nearby position $\mathbf{x}'$. The net result of these random moves of individual pollen grains is to cause the density of pollen grains to *diffuse* through space: if initially the pollen grains are concentrated at $\mathbf{x}$, over time they will spread out from $\mathbf{x}$ as they diffuse through space. In just the same way stars and dark-matter particles diffuse through action space.

Diffusion is particularly important for the stars of a stellar disc like the one we inhabit, because stars form in a very localized region of action space—the line associated with circular orbits in the disc's symmetry plane. As stars diffuse away from this line, their orbits become more eccentric and more highly inclined to the symmetry plane. Consequently, the random velocities of stars increase. Since the random velocities of gas molecules are associated with heat, we say the galactic disc 'heats'. The analogy is imperfect, however, since nothing heats the disc in the sense of supplying energy to it; it heats spontaneously and of its own accord, drawing the energy required for increased random velocities from its gravitational potential energy. The fluctuations in the gravitational field that cause disc heating are mostly generated by spiral structure (discussed below), but giant molecular clouds also contribute significantly. It is not clear whether infalling dwarf galaxies are significant contributors.

Globular clusters (Chapter 3, 'Globular star clusters') are very much like tiny galaxies. Their cores shrink and envelopes swell on account of the Poisson, $\sqrt{N}$, fluctuations in star density discussed above. In galaxies the timescale for Poisson fluctuations to cause measurable contraction in the core is much longer than the age of the Universe.

Destroying clusters   Most stars are born in small clusters— containing less than 1,000 $M_\odot$. Poisson fluctuations are large in such clusters and redistribute energy between the cluster's stars on a relatively short timescale. In any cluster a star requires only a finite amount of kinetic energy to escape completely from the cluster, and in any exchange of energy between stars there is always a chance that the gainer will acquire enough energy to escape. A star that does escape, never returns to risk losing energy. Since the energy of the cluster plus its escapees is conserved, the removal of positive energy by escapees must be mirrored by the energy of the surviving cluster becoming more negative. This is essentially the physics of evaporative cooling, which is what makes one shiver when wet in a draught.

As a cluster shrinks through evaporation of stars, Poisson fluctuations grow even more important and the rate of evaporative loss does not decrease even though the cluster's stars are on average becoming more tightly bound. Ultimately the cluster shrinks to a binary star; the energy released in the formation of this binary star has enabled every other star to escape to infinity.

We have just described what would happen to a small cluster if it were left alone for a long time. Actually, clusters are not isolated, but orbit through a galaxy and we will see on page 131 that they are gradually pulled apart by the galaxy's gravitational field. In fact, the Sun and every other star that is not now in a cluster has probably escaped from a cluster.

Spiral structure

Galaxies like ours contain spiral arms. In Chapter 4 we saw that the physics of accretion discs around stars and black holes is all about the outward transport of angular momentum, and that moving angular momentum outwards heats a disc. Outward transport of angular momentum is similarly important for galactic discs. In Chapter 4 we saw that in a gaseous accretion disc angular momentum is primarily transported by the magnetic field. In a stellar disc, this job has to be done by the gravitational field because stars only interact gravitationally. Spiral structure provides the gravitational field needed to transport angular momentum outwards.

In addition to carrying angular momentum out through the stellar disc, spiral arms regularly shock interstellar gas, causing it to become denser, and a fraction of it to collapse into new stars. For this reason, spiral structure is most easily traced in the distribution of young stars, especially massive, luminous stars, because all massive stars are young. In their short lives these stars don't stray far from their places of birth in an interstellar shock, so they trace the thin lines of the shocks. The gravitational field that caused the shock is mostly generated by numerous older, less massive stars. Their distribution forms a smoother spiral with broad arms, which is most prominent when a spiral galaxy is imaged in infrared light.

Spiral arms are waves of enhanced star density that propagate through a stellar disc rather as sound waves propagate through air. Like sound waves they carry energy, and this energy is eventually converted from the ordered form it takes in the wave to the kinetic energy of randomly moving stars. That is, spiral arms heat the stellar disc. Whereas sound waves heat air wherever they travel, spiral arms heat the disc at specific radii, where stars resonate with the wave. Exchange of energy between waves and particles at very specific locations is characteristic of collisionless systems, and is also crucial for the dynamics of electric plasmas. Our understanding of wave-particle interactions is still incomplete and the exact role they play in spiral structure and the evolution of galaxy morphology is controversial.

Origin of the bulge

Observations of the line-of-sight velocities of several thousand stars in the bulge/bar are well reproduced by bars that form in N-body simulations of self-gravitating discs. Bar formation proves to be a two-step process: quite a flat bar forms first, and a little later the bar experiences an instability which makes it vertically thicker. Thus the data are consistent with the proposition that all the stars of the bulge/bar and the disc were formed in the thin gas layer that occupies the equatorial plane. This accounts for the overwhelming majority of our Galaxy's stars: stars of the stellar halo were probably not formed in the Galactic plane, but less than 1 per cent of the Galaxy's stars belong to the stellar halo.

In N-body simulations of bars embedded in dark halos, the bar transfers angular momentum to the dark halo, with the consequence that the rate at which the figure of the bar rotates slows, and the bar grows stronger. Several pieces of evidence point to our bar having quite a fast rate of rotation, which is consistent with the N-body models if there has been a significant flux of gas through the disc and into the bar. In fact gas moves inwards because it loses angular momentum as it streams through spiral arms. As it approaches the bar's corotation radius, it starts to acquire angular momentum from the bar's rotating gravitational field, so its inward drift slows and its density increases—this is the origin of the *giant molecular ring*, a gas-rich region about 5 kpc in radius that surrounds the bar and dominates the Galaxy's star formation. Gas that escapes from the giant molecular ring and enters the bar rapidly loses angular momentum in shocks that we see in both numerical simulations and as dust lanes in external galaxies. After plunging rapidly through the bar, the gas builds up in a nuclear disc of radius ~ 0.2 kpc called the *central molecular zone*. There it feeds vigorous star formation as is evidenced by numerous supernova remnants in and near this disc.

Cannibalism

The Universe consists of dark-matter halos within which gravity has reversed the cosmic expansion, and our Galaxy is one of three substantial galaxies in one such halo, the Local Group. Because dwarf galaxies form in much greater abundance than giants, most of any halo's galaxies are dwarfs.

At the edge of any halo there are dwarf galaxies that are poised between their historic expansion from the halo's centre, and infall into the halo. These objects are on highly eccentric orbits in the halo's gravitational field and are just now at their most distant from the halo's centre. Several gigayears from now, they will come quite close to the centre of the halo. How will this experience affect them?

As a galaxy falls into a halo from r_{apo}, it attracts dark-matter particles that are near its path. To grasp the consequences of this attraction, it's easiest to imagine that the dark-matter particles constitute a fluid. As the dwarf passes, the fluid receives an impulse towards its line of flight and converges on that line (Figure 37). However it takes time for the fluid to move and its density to be enhanced along the line of flight. So the region of enhanced dark-matter density lies behind the dwarf. Hence the gravitational attraction from the region of enhanced density pulls the dwarf backwards; gravity acts to retard the dwarf's motion as if it were experiencing friction. In fact this phenomenon is called *dynamical friction*.

Since the impulse from the dwarf that creates the region of dark-matter overdensity is proportional to the mass of the dwarf, the mass of the overdensity is proportional to the mass of the dwarf. The drag on the dwarf, being proportional to the product of the masses of the dwarf and the overdensity is thus proportional to the *square* of the dwarf's mass, and the dwarf's deceleration is

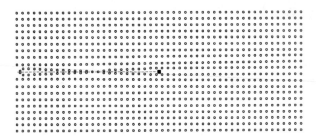

37. Formation of a wake behind a massive body. The massive body marked by the central black square is moving from left to right. Particles initially stationary at the nodes of a regular grid move towards it. A region of enhanced density of these particles is evident along the massive body's past track.

proportional to the dwarf's mass. Hence the more massive an infalling galaxy is, the more rapidly dynamical friction modifies its orbit.

If the dwarf did not experience dynamical friction, it would pass near the centre of the halo and move back out to nearly the radius from which it started; it wouldn't actually get to r_{apo} because other infalling material would have increased the mass interior to r_{apo} since it was last there and it wouldn't have enough energy to reach r_{apo}. If the dwarf has experienced much dynamical friction, it will stop moving out long before it reaches r_{apo}. Dynamical friction will cause each radius of turnaround to be smaller than the preceding one, and ultimately by this reckoning the dwarf will end up at rest in the centre of the halo. The halo, or the galaxy at its centre, will have cannibalized the dwarf.

If events unfolded precisely as just described, a dwarf of mass m that fell into a halo of mass $M(r)$ would survive for a time of order

$$t_{eat} = \frac{M(r_{apo})}{m} T,$$

where T is the period of a circular orbit at radius r_{apo}. If we extrapolate this formula to the case of a merger of equals, $m \simeq M(r)$, such as the future encounter of our Galaxy with the Andromeda Nebula, M31, we find $t_{eat} = T$, the initial orbital period. N-body simulations show that this extrapolated result is actually correct.

The dwarf that is currently closest to our Galaxy is the Sagittarius Dwarf, which lies ~ 13 kpc from the Galactic Centre on the side opposite the Sun. The period T of a circular orbit at this radius is $T \simeq 0.5$ Gyr or $\sim 1/25$ of the current age of the Universe. Hence for t_{eat} to be smaller than the current age of the Universe, the Dwarf's mass must exceed 10^{10} M$_\odot$ since $M(20\,\text{kpc}) \simeq 2.5 \times 10^{11}$. The mass of the Dwarf is poorly determined, but it is much smaller than 10^{10} M$_\odot$, so we might conclude that it isn't going to be eaten any time soon. Not so! For we have underestimated our Galaxy's digestive powers.

Washed away on the tide

When an extended body orbits in a gravitational field, it is stretched along the line that joins its centre to the centre of attraction. The reason is that material on its surface that is closest to the centre of attraction feels a stronger gravitational force than material that is on the far side from the centre of attraction (Figure 38). Hence these two bodies of material want to accelerate at different rates. But they are parts of a single body, so they are obliged to accelerate at a single, compromise rate. This rate is too small for the material on the side of the centre of attraction, and too fast for the opposite material. The body responds to this discord by stretching along the line of centres so its gravitational field pulls back material nearest the centre of attraction, and urges on material on the opposite side. Since the physics we have described is how the Moon raises tides on the surface of the oceans, we say bodies are stretched by the *tidal field* of the body they are orbiting.

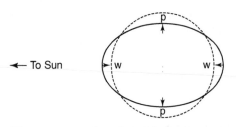

38. Tides without a moon: at the points of the Earth nearest to and furthest from the Sun the surface of the ocean (full curve) is higher than in the mean (shown dotted), so pressure is too small to balance the gravitational attraction of the Earth and net force *w* on a small volume of water is downwards. On the left this net force cancels a part of the gravitational attraction of the Sun, while on the right it adds to the Sun's attraction. These additions to the attraction of the Sun ensure that elements of the ocean at both locations orbit the Sun at the same angular velocity.

As dwarfs move inwards, their dark-matter halos are stretched to the point that particles are pulled clean out of the dwarf's clutches and start orbiting independently through the Galaxy's gravitational field. Particles that leave the dwarf on the same side as the centre of attraction transfer to orbits that have less angular momentum than the dwarf, and pull ahead of it, whereas particles that leave on the far side transfer to orbits of greater angular momentum and fall behind. In this way two *tidal tails* form and the dwarf grows less massive (Figure 39).

Now that the dwarf has become less massive, its gravitational field is weaker, and the points at which particles break free of the dwarf and start orbiting independently, edge towards the dwarf's centre. A vicious circle of mass loss proceeds, with the dwarf becoming less and less massive and the tidal tails growing longer and longer. At some stage the points at which particles break free come close enough to the centre that large numbers of stars as well as dark-matter particles stream out into the tidal tails. The Sagittarius dwarf reached this stage some time ago, and tidal tails

39. A simulation of tidal tails forming. The curve shows the orbit of the centre of the cluster that is being ripped apart.

containing its stars now wrap around the Galaxy at least once and perhaps twice.

By measuring the colours and brightnesses of millions of stars, it has been possible to examine the density of stars in spherical shells centred on the Sun. The density proves to be full of ridges and overdense patches. In fact, the ridges and overdense patches contain at least half the stars of the stellar halo. So it is probable that the stellar halo is entirely built up by tidal streams stripped from dwarf galaxies and globular clusters. When some of these objects first fell into our Galaxy, they may have been sufficiently massive to experience significant dynamical friction. But as they orbited, tides caused them to lose more and more of their dark-matter halos, and at some point their masses fell below the threshold for significant dynamical friction. However, tidal stripping continued relentlessly, and they were eventually completely digested without reaching the Galactic centre.

Chemical evolution

Nearly all the elements heavier than lithium have been produced since the Galaxy formed, and while they were formed, star formation has continued in the Galaxy. So in stars of various ages we have a fossil record of how the heavy-element content of the interstellar medium has evolved. Moreover, since old stars tend to have larger random velocities, the chemical content of a star is correlated with its kinematics.

It is hard to determine the age of an individual star because to do so one needs precise values for its mass, luminosity, and chemical content, and the mass is especially hard to determine. So astronomers like to use chemical content, which is easier to measure, as a surrogate for age.

Supernovae are the most important contributors to the enrichment of the interstellar medium with heavy elements. In Chapter 3, 'Exploding stars', we saw that there are two fundamentally different kinds of supernova: core collapse supernovae, which mark the deaths of massive stars ($M > 8\,M_\odot$), and deflagration supernovae, which occur when a white dwarf accretes too much material from a companion. Since massive stars have short lives, a burst of core-collapse supernovae follows soon after ($\sim 10\,\text{Myr}$) an episode of star formation, while evolution to a white dwarf followed by significant accretion probably takes $\sim 1\,\text{Gyr}$. So for the first gigayear of the Galaxy's life only core-collapse supernovae enriched the interstellar medium.

As we saw in Chapter 3, 'Exploding stars', deflagration supernovae produce mostly iron while core-collapse supernovae produce a wide spectrum of heavy elements. It follows that the abundance in the interstellar medium of iron relative to say magnesium or calcium would have been lower in the first gigayear of the Galaxy's life than it is now. Figure 40 shows that one can identify two populations of stars near the Sun: in the *alpha-enhanced* population the abundance of Mg and Ca is higher at a given value of the Fe abundance than it is in the *normal-abundance* population. We infer that stars of the alpha-enhanced population were formed in the first $\sim 1\,\text{Gyr}$ of the galaxy's life.

Within each population a wide range of abundances of Fe relative to hydrogen occurs. One possibility is that all stars with low values

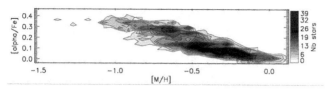

40. Stars formed in the first gigayear have low abundances of Fe relative to Mg and lie high up in this diagram. The horizontal coordinate is the abundance of iron. The contours show the density of stars in this plane.

of Fe/H formed before stars with higher values. But this conclusion is in general false because it is likely that early on interstellar gas was converted into stars more rapidly towards the centre of the Galaxy than further out, with the consequence that the heavy-element abundance rose near the centre much more rapidly than further out. Hence a given value of Fe/H was achieved earlier at small radii than at large radii, and a star with low Fe/H might have formed near the Galactic centre early in the life of the Galaxy, or more recently at large radii.

A star's ratio value of Mg/Fe provides a degree of discrimination between these possibilities: if Mg/Fe is high, the star must have formed in the first gigayear of the galaxy's life, and at that time only low values of Fe/H would have been attained at large radii, so a star with high Mg/Fe and high or moderate Fe/H must have formed quite far in.

Alpha-enhanced stars have large random velocities and extend further from the Galactic plane than normal-abundance stars. The data are consistent with the thick disc comprising alpha-enhanced stars and the thin disc normal-abundance stars.

Given the tight connections between time and place of birth, chemical composition, and present kinematics of disc stars, it is very useful to model chemical and dynamical evolution together. In such a model stars are born at some rate at each radius from the local interstellar gas, and enrich this gas with Mg and Ca shortly after they are formed, and with Fe a gigayear or so later. Stars are born on nearly circular orbits and gradually wander to more eccentric and inclined orbits. The interstellar gas at each radius is depleted by star formation and the ejection as a galactic wind of gas heated by supernovae. It is augmented by accretion of intergalactic gas. Spiral structure drives the star-forming gas slowly inwards, carrying heavy elements with it. The goal of such a model is to reproduce the observed kinematics of stars as a function of position, and the observed correlations between chemical composition and kinematics. This is currently an active area of research.

The great reservoir

In Chapter 2, 'The gas disc', we described our Galaxy's gas disc, which is typical of the gas discs of spiral galaxies and contains $\sim 6 \times 10^9 \, M_\odot$. Since the Galaxy turns gas into stars at a rate of about $2 \, M_\odot \, \mathrm{yr}^{-1}$, this stock of material for star formation will be exhausted within $\sim 3 \, \mathrm{Gyr}$. Is the current gas disc the small remnant of a massive gas disc from which the Galaxy has formed $\sim 5 \times 10^{10} \, M_\odot$ of stars, or is it just a buffer between star formation and gas accretion?

Observations of other galaxies show that the star-formation rate in a disc is proportional to the surface density of cold gas, so in the absence of accretion, the mass of the gas disc decreases exponentially with time. Given the rate at which our Galaxy is currently forming stars and it current gas mass, it is easy to show that in the absence of accretion the gas mass 10 Gyr in the past would have been $\sim 1.7 \times 10^{11} \, M_\odot$, and nearly all this mass would now be in stars. This absurd conclusion shows that our premise, that the Galaxy does not accrete gas, is false.

From the cosmic microwave background (CMB) we can read off the current cosmic mean density of ordinary matter. By measuring the luminosities of galaxies we know the cosmic mean luminosity density, so we can determine what mass of ordinary matter is required, on the average, to generate a given amount of luminosity – ~ 40 $M_\odot/L_\odot$. Now if you take any reasonably representative group of galaxies, from the group's luminosity, you can deduce the quantity of ordinary matter it should contain. This quantity proves to be roughly ten times the amount of ordinary matter that's in the galaxies. So most ordinary matter must lie *between* the galaxies rather than within them.

Intergalactic atomic hydrogen can be detected by its absorption of *Lyman alpha* photons that come to us from distant quasars (Chapter 4, 'Quasars'). Since light from the most distant quasars has taken most of the life of the Universe to reach us, it has sampled intergalactic space at essentially every cosmic epoch. Hence, measurements of the Lyman alpha line enable us to track the cosmic density of hydrogen over time. In the first gigayear, the density of hydrogen was roughly the same as the density of ordinary matter inferred from the CMB, but as time went on the density of hydrogen fell and now it is less than 1 per cent of the expected mean density of ordinary matter. Searches for 21 cm emission from atomic hydrogen between nearby galaxies confirms that there is very little intergalactic hydrogen now.

The natural interpretation of this finding is that the cosmic stock of hydrogen has been used up to create stars and galaxies, but studies of nearby galaxies show that they don't contain sufficient ordinary matter to make this a viable explanation. The accepted explanation is that the missing matter *is* in intergalactic space, but it so hot that it is completely ionized and can therefore not be detected with any spectral line of hydrogen.

Actually, it is natural for intergalactic gas to be extremely hot because the pressure within the gas can resist the gravitational pull of galaxies only if the temperature of the gas exceeds the *virial temperature*, which is the temperature at which the thermal velocities of atoms are comparable to the orbital speeds of dark-matter particles – in our Galaxy this temperature is $\sim 2 \times 10^6 \, \mathrm{K}$ but it varies as you move around the Universe, approaching $10^8 \, \mathrm{K}$ in the richest clusters of galaxies. At the virial temperature the pressure in the gas provides an effective counterbalance to gravity, so the density of the gas tends to track the density of dark matter. Gas in which atoms move much slower than the dark-matter particles has too little pressure to resist gravity, and in equilibrium must be confined to a thin rotating disc.

Gas at the virial temperature emits X-rays. In rich clusters of galaxies the emitted X-rays are strong enough to be detected and indicate that the predicted quantity of gas is present (Figure 41). Outside rich clusters the X-ray emission is expected to be too weak and too soft to be detected by current telescopes. However, hints of the presence of the expected gas are found in the ultraviolet spectra of some objects.

Looking for absorption lines in the spectrum of a background object is by far the most sensitive way to search for gas (page 137).

Astrophysics

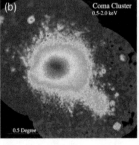

41. The Coma cluster, left (a) in optical light and right (b) in X-rays. The optical picture is only 0.32° wide while the X-ray image is 2.7° wide, so the optical picture shows only the centre of the cluster. The bright object at the top and to the right of centre in the optical picture is a star in our Galaxy; all other objects are galaxies.

The best waveband to examine for absorption lines depends on the temperature of the gas you are seeking, because the waveband should include photons that have the right energy to lift an ion out of its ground state into an excited state, and at high temperatures only tightly bound electrons remain bound to ions, and the ions can only be excited by energetic photons.

The ideal waveband to search for gas at more than 10^6 K is the X-ray band. Unfortunately X-ray telescopes are hard to build because X-rays tend to knock electrons out of mirrors rather than being reflected by them. Moreover for a given luminosity the rate of emission of X-ray photons is a thousand times less than the rate of emission of optical photons, so X-ray photons are scarce and statistical noise is a big problem. For these reasons sensitive searches for X-ray absorption lines are not possible. The Hubble Space Telescope has searched for absorption lines in ultraviolet spectra, however. Lines indicating the presence of ions such as five times ionized oxygen, O^{5+}, and three times ionized carbon, C^{3+} are detected in the spectra of quasars. The velocities of these lines indicate that the absorption occurs where the line of sight passes by a galaxy, but often at a great distance, typically 100 kpc.

The detected ions are common in gas that is cooler ($\sim 3 \times 10^5$ K) than we think the bulk of the gas is, and the usual interpretation of these data is that we are seeing absorption at the interface between the gas that fills most of intergalactic space and small clouds of much cooler ($\sim 10^4$ K) gas. This is a very active area of research and our interpretation of the data may be different in a few years.

Drivers of morphology

The nature of a galaxy is largely determined by three numbers: its luminosity, its bulge-to-disc ratio, and the ratio of its mass of cold gas to the mass in stars. Since stars form from cold gas, this last ratio determines how youthful the galaxy's stellar population is.

A youthful stellar population contains massive stars, which are short-lived, luminous, and blue (Figure 3). An old stellar population contains only low-mass, faint, and red stars. Moreover, the spatial distribution of young stars can be very lumpy because the stars have not had time to be spread around the system, just as cream just poured into coffee is distributed in blobs and streamers that quickly disappear when the cup is stirred. Old stars, by contrast, are smoothly distributed. Hence a galaxy with a young stellar population looks very different from one with an old population: it is more lumpy/streaky, bluer, and has a higher luminosity than a galaxy of similar stellar mass with an old stellar population.

The ratio of bulge-to-disc mass obviously affects the shape of the galaxy, especially when the disc is viewed edge-on. It also affects the compactness of the galaxy because a bulge of a given luminosity tends to be more compact than the corresponding disc.

Finally, the structure of a galaxy is profoundly affected by its luminosity, because the latter is related to its stellar mass, which is in turn connected to the speeds at which its stars and dark-matter particles move. Hence luminosity is connected to its virial temperature. A massive, luminous galaxy has a high virial temperature, which makes it hard for supernovae to drive gas out of the galaxy. Conversely, it is easy for supernovae to drive gas out of a low-mass, low-luminosity galaxy. Since today's cold gas is tomorrow's stars, driving gas out of a galaxy early on depresses the ratio of stars to dark matter, and this is thought to be the reason why low-luminosity galaxies have high ratios of dark-matter mass to stellar mass.

Galaxies are also shaped by their environments. Dense environments are rich in elliptical and lenticular galaxies, while abnormally under-dense environments are rich in dwarf irregular galaxies. Spiral galaxies like our own tend to inhabit regions of intermediate density. Such regions are made up of patches about a megaparsec in diameter within which the cosmic expansion has been reversed by gravity, so now each region consists of a number of galaxies that are falling towards one another. Our Galaxy lies in just such a region, that of the Local Group of galaxies. A few galaxies within such a group are stationary with respect to the local intergalactic gas, and act as sinks for this gas as it gradually cools. By processes that are still not fully understood but probably involve dynamical interactions with clouds of cold gas shot up out of the disc by supernovae (Chapter 2, 'The gas disc'), virial-temperature gas cools onto the disc, augmenting the supply there of cold, star-forming gas. Hence these galaxies, which include the Milky Way and our neighbours M31 and M33, remain youthful.

Smaller galaxies orbit around and through each of these major galaxies. Such galaxies cannot replenish their cold gas by accreting virial-temperature gas because they are moving through that gas too fast. Hence star formation in these galaxies dies out when whatever initial stock of cold gas they had is exhausted. This is why the nearer a satellite galaxy lies to the centre of its host galaxy, the less likely we are to see it forming stars: the density of virial-temperature gas increases inwards, so galaxies that are close in experience the strongest winds as they move at roughly the sound speed through the virial-temperature gas. These strong winds (think of travelling in an open-topped Boeing 747) sweep the satellite's own gas away (Figure 42).

Galaxy clusters The densest environments are rich clusters of galaxies. These are regions about a few megaparsecs in size in which gravity reversed the cosmic expansion quite long ago, so the density of galaxies and the virial-temperature are both high. Because the virial-temperature gas is dense, its X-ray emission is rather intense and can be detected by X-ray telescopes out to significant distances from cluster centres (Figure 41). The virial-temperature gas of a cluster is unusually hot because the characteristic velocities and temperatures of cosmic structures

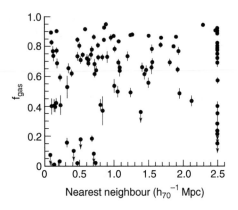

42. The further a dwarf galaxy is from its neighbours, the more likely it is to contain little cold, star forming gas.

increases with mass scale, and a rich cluster of galaxies is extremely massive ($\sim 10^{15}$ M$_\odot$).

The virial temperatures of rich clusters exceed the temperature to which supernovae can heat interstellar gas. Hence no ordinary matter can have been blown out of these regions since the Big Bang, and their ratios of the mass of ordinary matter to dark matter should be the cosmic ratio. Within the experimental errors this is found to be the case.

Galaxies that orbit within the cluster cannot acquire intergalactic gas through cooling, so they are rarely forming significant numbers of stars. Galaxies that have cold, star-forming gas discs continually fall into rich clusters, and they go on forming stars at a declining rate until their stock of cold gas is exhausted. During this phase these are called *anaemic spiral galaxies*. Once they have ceased forming stars, anaemic spirals become lenticular galaxies—their stellar discs are fossil relics of their former star-forming gas discs.

Usually (but not always), the centre of a rich cluster has an exceptionally massive galaxy at its centre. Such a *cluster-dominant galaxy* is rather a special beast in that it is at rest with respect to the local intergalactic virial-temperature gas, so it can accrete gas and we might expect it to have a significant rate of star formation. Some of these galaxies, for instance NGC 1275 at the centre of the Perseus galaxy cluster, do have significant numbers of young stars, but most do not, and those that do lack dominant stellar discs.

Why do cluster-dominant galaxies not develop giant stellar discs that would make the parent galaxy a hugely scaled-up version of a spiral galaxy such as ours, in which the observed cluster-dominant galaxy would be the bulge and the cluster a halo of satellites? Astrophysicists do not have a complete answer to this question but the final answer will surely involve two key pieces of physics. First, the way a plasma cools changes fundamentally at around 10^6K because the common elements carbon and oxygen are stripped of their last electrons around this temperature. Even a very low density of ions that have bound electrons greatly increases the cooling capacity of a plasma because a bound electron radiates photons *very* much more efficiently than a free electron. This phenomenon causes the time required for plasma of a given density to cool to rise steeply as the temperature increases from 10^6K to 10^7K, and this will hinder the formation of a disc of cool, star-forming gas around a cluster-dominant galaxy.

The second key fact is the presence of super-massive black holes (masses $\sim 10^9$ M$_\odot$) at the centres of most luminous galaxies. Any virial-temperature gas that does cool will do so near the black hole at the centre of the cluster-dominant galaxy, where the gas is under the most pressure and therefore densest. When the black hole accretes some of this gas, jets form (Chapter 4, 'Jets'), which blast through the surrounding virial-temperature gas, reheating it. Radio-frequency and X-ray observations of cluster-dominant galaxies provide clear evidence of this process. In particular, the X-ray spectra of several clusters show that the gas there is up to three times cooler than gas in the main body of the cluster, but the gas is not in the process of cooling right down. This finding is a clear sign that re-heating prevents the centre from settling into a steady state in which gas flows steadily onto the central body.

Our own Galaxy has a much smaller central black hole than a typical cluster-dominant galaxy: its mass is $4 \times 10^6 \, M_\odot$. Nonetheless, it has the capacity to re-heat the plasma that envelops it, and it probably regularly does so. At the moment it appears to be resting, like a dormant volcano. Its activity explains why the centre of our Galaxy is not the region of most intense star formation—that honour belongs to the central molecular zone. As we saw in our discussion of the Galactic bulge, cold gas is fed into this ~ 0.2 kpc radius zone from the ~ 5 kpc radius giant molecular ring. The key to understanding why every cluster of galaxies is not dominated by a giant spiral galaxy lies in understanding why these clusters have no analogues of the giant molecular ring, and this will likely be traced to the change around temperature 10^6 K in the way a plasma cools.

Chapter 8
The big picture

In Chapter 6 we discussed a number of phenomena that can be explained with the general theory of relativity. However, by far the biggest contribution of general relativity to astrophysics was to make it possible to discus the geometry and dynamics of the entire universe—it made cosmology a branch of physics rather than of philosophy or theology. We do not have space here for a systematic account of cosmology—for that the reader can turn to the *Cosmology: A Very short Introduction*. Instead we outline our current understanding of how stars and galaxies emerged from the big bang, in this way providing some context for the physical processes introduced in preceding chapters.

At heart cosmology is about three fluids: *dark energy*, which nobody understands; dark matter, which nobody can see; and the cosmic microwave radiation, which dominated the universe prior to a redshift $z \sim 3{,}000$ and can be studied in great detail as it constitutes the CMB. Dark energy became dominant rather recently ($z \sim 0.5$) and the intervening era was dominated by dark matter. From the cosmological standpoint what distinguishes these three fluids is the pressure they exert. Radiation exerts a positive pressure, dark matter exerts negligible pressure, and dark energy exerts negative pressure: that is it exerts tension.

According to general relativity, pressure is a source of gravitational attraction just as much as mass-energy. Hence the gravitational pull of the Sun on the Earth is larger than it otherwise would be because the pressure deep inside the Sun is high. Conversely, tension generates gravitational repulsion, and in the case of dark energy the repulsion generated by its tension overwhelms the attraction generated by its energy density. Since dark energy now dominates the universe, the latter is now being blasted apart by the gravitational repulsion that it generates. That this is happening was discovered by measuring the redshifts and distances of deflagration supernovae (Chapter 3, 'Exploding stars'). From these measurements the rate at which the universe was expanding at past epochs has been inferred, and it seems that around $z \sim 0.5$ the expansion rate started to increase, whereas previously it had decreased with time.

For the first 200,000 years after the big bang the universe was very nearly homogeneous and dominated by radiation, so gravity was strongly attractive and continuously slowed the expansion of the cosmic fireball. Because it exerts pressure, the radiation fluid did work on the expansion, and on account of doing this work its energy density diminished faster than the energy density of dark matter, which did negligible work because it exerted negligible pressure. Consequently, at redshift $z \sim 3,000$ the energy density of radiation fell below that of dark matter. At this point the initially tiny fluctuations in energy density with position began to grow at a significant rate, because the gravitational field was now predominantly generated by dark matter rather than radiation, and pressure did not oppose the tendency for some regions to become more dense than others. At this stage ordinary matter was tightly locked to the nearly homogeneous radiation fluid, so it did not participate in the clustering of dark matter. Then at a redshift $z \sim 1,000$ the temperature of the radiation dropped to the value at which electrons became bound to protons and alpha particles, to form hydrogen and helium atoms. The formation of these atoms effectively decoupled ordinary matter from the radiation fluid because the atoms scarcely scattered photons. Now nothing resisted the gravitational pull of ordinary matter into regions of enhanced dark-matter density, and the formation of structure got underway in earnest. We can study the *epoch of decoupling* in great detail by measuring the properties of the CMB because its constituent photons have travelled to us unmolested since the epoch of decoupling.

At the epoch of decoupling, the fluctuations in the density of dark matter were only parts in 100,000 so it took a long time for gravity to amplify them sufficiently for the regions of highest density to cease expanding and to collapse into stars and galaxies. This started to happen around redshift $z \sim 15$, and the first collapsed objects included massive stars. These stars radiated energetic photons that gradually re-ionized the hydrogen and helium atoms, a process that was largely complete by $z \sim 6$.

Currently we don't have much observational data relating to the redshift interval from $z \sim 1,000$ to 6, but from redshift 6 the observational record is significant. A few very massive galaxies must have already formed by then because luminous quasars are known at $z > 7$, and we know (Chapter 4, 'Quasars') that these are powered massive black holes that sit at the centres of massive galaxies.

Although at $z \sim 6$ there were a few massive galaxies, only a tiny fraction of the present-day stars had formed. At that time most galaxies were much smaller than present-day galaxies, and there was an abundance of cool, dense gas. As this gas flowed into nascent galaxies, it formed stars at a rate that continued to increase up to redshift $z \sim 2$. The flow of gas into galaxies was chaotic, so often the gas was not organized into a thin, flat disc like our Galaxy's present gas disc. Instead streams of gas raced hither and thither, crashing into each other and rapidly forming stars when they did so.

Massive black holes were in the thick of the melé, hoovering up gas as fast as they could. So galactic bulges and black holes grew fast at this time. Energy released by accretion onto the black holes was converted by the abundant ambient gas into optical and infrared photons, causing the region around each black hole to shine brightly as a quasar. Energy released at the deaths of massive stars heated the surrounding interstellar gas (Chapter 2, 'The gas disc'), with the consequence that an ever increasing fraction of the volume in and around galaxies became occupied by gas at the virial temperature or above. Gas hotter than the virial temperature flowed out into intergalactic space, carrying with it much of the heavy elements that had been synthesized by the recently deceased stars (Chapter 7, 'Drivers of morphology').

From redshift $z \sim 2$ the rate of star formation and black hole feeding gradually diminished as the flow of gas onto galaxies slackened, and more of the gas became too hot to form stars or to allow a black hole to gorge itself. In Chapter 7, 'Drivers of morphology', we described how these global trends influenced the morphologies of individual galaxies.

So there's a very brief history of the universe. Much of the physics involved is extremely complex and we are far from understanding how the various processes played out. Consequently, if we were to go into much more detail, we would soon reach the limits of our current understanding.

The universe is a huge canvas, and nature has wrought on it with very many techniques. Our knowledge of the canvas and of the artist's methods is growing rapidly, but we have much, much more to learn.

Further reading

Chapter 2: Gas between the stars

Bruce Draine, *Physics of the Interstellar and intergalactic Medium* (Princeton University Press, 2011). (A definitive, graduate-level text.)

T. Padmanabhan, *Theoretical Astrophysics*. Vol II: *Stars and Stellar Systems* (Cambridge University Press, 2001).

Chapter 3: Stars

Andrew King, *Stars: A Very Short Introduction* (Oxford University Press 2012).

T. Padmanabhan, *Theoretical Astrophysics*. Vol II: *Stars and Stellar Systems* (Cambridge University Press, 2001).

Dina Prialnik, *The Theory of Stellar Structure and Evolution* (Cambridge University Press, 2009). (A lucid, undergraduate-level text.)

A. Sarajedini, L.R. Bedin, B. Chaboyer, et al., The ACS Survey of Galactic Globular Clusters. I. Overview and Clusters without Previous Hubble Space Telescope Photometry, *The Astronomical Journal* 133, 4 (2007), 1658–72.

Chapter 4: Accretion

K.M. Blundell and M.G. Bowler, Letters, *Astrophysical Journal* 616 (2004), L159.

Juhan Frank and Andrew King, *Accretion Power in Astrophysics* (Cambridge University Press, 2002).

T. Padmanabhan, *Theoretical Astrophysics*. Vol II: *Stars and Stellar Systems* (Cambridge University Press, 2001).

Chapter 5: Planetary systems

Stephen Eales, *Planets and Planetary Systems* (Wiley-Blackwell, 2009).

W. Kley and R.P. Nelson, Annual Reviews, *Astronomy and Astrophysics*, 50 (2012), 211–49.

Chapter 6: Relativistic astrophysics

D.P. Bennett, S.H. Rhie, S. Nikolaev, et al., Masses and Orbital Constraints for the OGLE-2006-BLG-109Lb,c Jupiter/Saturn Analog Planetary System, *Astrophysics Journal* 713 (2010), 837–55.

M. Krause, Very Light Jets II: Bipolar Large Scale Simulations in King Atmospheres, *Astronomy and Astrophysics*, 431 (2005), 45–64.

Maurice van Putten and Amir Levison, *Relativistic Astrophysics of the Transient Universe* (Cambridge University Press, 2012).

Albert Einstein, *Relativity—The Special and General Theory* (Forgotten Books, 2015). (Get the story from the horse's mouth—this book is based on lectures Einstein delivered to engineers.)

P.A.M. Dirac, *General Theory of Relativity* (Princeton University Press, 2011). (A wonderfully short and elegant exposition by perhaps the second-greatest theoretical physicist of the 20th century.)

Chapter 7: Galaxies

James Binney and Michael Merrifield, *Galactic Astronomy* (Princeton University Press, 1998). (Graduate-level text.)

Bruce Draine, *Physics of the Interstellar and Intergalactic Medium* (Princeton University Press, 2011). (A definitive, graduate-level text.)

M. Geha, R. Blanton, and M. Masjedi, The Baryon Content of Extremely Low Mass Dwarf Galaxies, *The Astrophysical Journal*, 653 (2006), 240–54.

D.H. Jones, B.A. Peterson, M. Colless, and W. Saunders, Near-Infrared and Optical Luminosity Functions from the 6dF Galaxy Survey, *Monthly Notices of the Royal Astronomical Society* 369 (2006), 25–42.

A. Recio-Blanco, P. de Laverny, G. Kordopatis, et al., The Gaia-ESO Survey: The Glactic Thick to Thin Disc Transition, *Astronomy and Astrophysics* 567, id.A5 (2014), 21 pp.

J.L. Sanders and J. Binney, Stream-Orbit Misalignment I: The Dangers of Orbit-Fitting, *Monthly Notices of the Royal Astronomical Society* 433, 3 (2013), 1813–25.

L. Sparke and J.S. Gallagher, *Galaxies in the Universe* (Cambridge University Press, 2007). (Undergraduate-level text.)

Further reading

Index

Astrophysics

Astrophysics